T0201699

# BIOLOGY
# 100 IDEAS IN 100 WORDS

**A whistle-stop tour of key concepts**

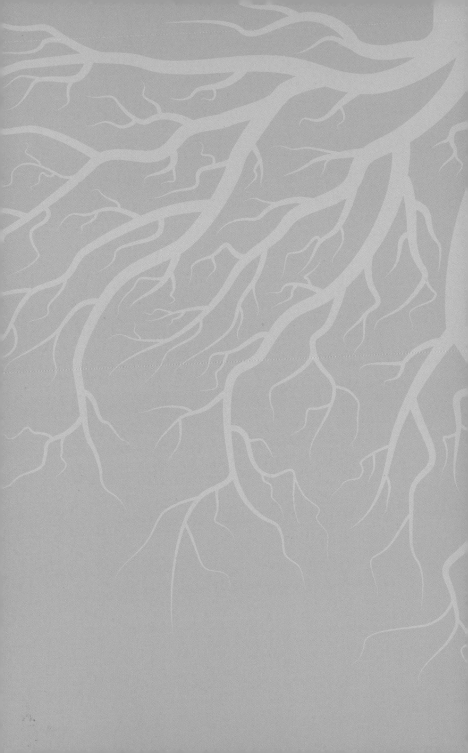

SCIENCE
MUSEUM

# BIOLOGY
# 100 IDEAS IN 100 WORDS

## A whistle-stop tour of key concepts

Eva Amsen

# Contents

# Introduction

People have studied the living world for thousands of years. Early humans learned how to cultivate plants and breed animals for food. They explored human health and how to treat illnesses. And they studied how different organisms all found a place in nature. A lot of what we learned came from careful observation. That's how we know that plants grow better when they get enough sun and water, that humans get sick if they eat rotten food and that some animals lay eggs.

Initially, all of these observations were based on things that people could see with their own eyes. But that didn't stop early biologists from making impressive discoveries. For example, physician Ibn Sina (sometimes known as Avicenna) described cancer in the early 11th century, even though he didn't yet understand what caused it. And the 17th-century biologist Maria Sybilla Merian created detailed illustrations that clearly showed how caterpillars turned into butterflies.

There's only so much you can observe by eye, though. Many modern discoveries in biology rely on technology and one of the most important technological developments for biologists was the invention of the microscope in the 17th century. Robert Hooke summarised this new exciting microscopic world in his 1665 book *Micrographia* by writing that "by the help of Microscopes, there is nothing so small as to escape our inquiry; hence there is a new visible World discovered to the understanding."

He was right: much of today's research in cell biology, genetics, stem cell research, immunology, microbiology and other fields of biology would not be possible without microscopes that are getting increasingly better at seeing even the tiniest details of life.

Another boost to understanding biology was learning which molecules are important to biology. Many of those discoveries are from the 19th and 20th century, during a time when chemists studied molecules of the living world such as ATP, lipids, glucose, DNA and proteins.

Biologists are still making many new discoveries with ever-improving technologies. Since the last few decades it has become a lot cheaper and easier to explore the genetic code of any type of organism. That means we can study biology in entirely new ways, but it also produces a lot of new information. To make sense of that all that data, software analysis and computational models are now just as much a part of biology research as the microscope has been for the last four centuries.

All those discoveries also created new ways to use knowledge from biology. Researchers are finding ways to protect crops by studying genetic diversity, treating cancer by editing cells of the immune system or using computational data to predict how ecosystems adapt to change. So even though the methods may have changed over time, biologists are still driven by the same goals as thousands of years ago: providing food, keeping people healthy and protecting the diversity of the world around us.

# Classification

Biology is the study of all of life. That includes millions of different kinds of animals, plants and smaller organisms like bacteria.

Biologists have agreed on a system to group all organisms into increasingly specific categories. Each organism also has an official Latin name based on their genus and species categories. This classification system, or taxonomy, makes it easier to compare organisms and to talk about their differences and similarities.

Understanding what sets organisms apart as unique species has also helped biologists to research evolution, genetics and developmental biology.

# 1

# Taxonomy

**WHY IT MATTERS**
Grouping and naming organisms makes it easier to talk about the things they have in common, such as their physiology or habitat

**KEY THINKERS**
Mayans and ancient Egyptians (c. 1500 BCE) grouped plants and animals. Aristotle (384–322 BCE) documented a detailed taxonomy. Carl Linnaeus (1707–1778) introduced the naming system. Charles Darwin (1809–1882) showed related species on a tree

**WHAT COMES NEXT**
The classification of organisms is a work in progress. Biologists now use genetic sequencing to identify new species and determine how they are related

**SEE ALSO**

Biologists study millions of different **organisms**, from the smallest **microbe** to the largest whale. They group all of these organisms into categories to make them easier to discuss.

Larger categories are divided into smaller ones, down to the level of individual **species**. Each organism has a Latin name based on its genus and species.

Initially, taxonomy was based on what organisms looked like, but now biologists mostly use **genetic** information to distinguish different species in more detail and show how they are related. This can be shown visually on a tree of life, or phylogenetic tree.

TAXONOMY OF LIVING THINGS

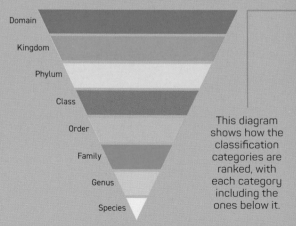

This diagram shows how the classification categories are ranked, with each category including the ones below it.

# In 100 words

Taxonomy in biology is the organization of all organisms into specific categories. All living things are part of one of three domains: Archaea, Bacteria or Eukarya. Within each domain are kingdoms, such as Animalia (animals) within Eukarya. These are further divided into phylum, class, order, family, genus and species.

Every organism has a unique Latin name consisting of the name of the genus and the species to which it belongs. Humans are *Homo sapiens*. Originally, taxonomy was based on how organisms looked and behaved, but modern biology uses evolutionary origins and genetics to place species on the tree of life.

A phylogenetic tree such as this one shows how similar genetic material is between related organism groups. Animals are most similar to the close branches of plants and fungi.

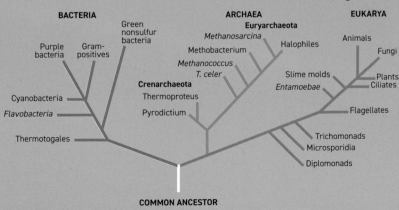

**BACTERIA**

Green nonsulfur bacteria

Purple bacteria

Gram-positives

Cyanobacteria

*Flavobacteria*

Thermotogales

**ARCHAEA**

**Euryarchaeota**

*Methanosarcina*

Methobacterium

*Methanococcus*

*T. celer*

Halophiles

**Crenarchaeota**

Thermoproteus

Pyrodictium

**EUKARYA**

Animals

Fungi

Slime molds

*Entamoebae*

Plants

Ciliates

Flagellates

Trichomonads

Microsporidia

Diplomonads

**COMMON ANCESTOR**

11

# 2

# Prokaryotes

If you did a head count of all the individual **organisms** on Earth, most of them would be members of the domains Bacteria or Archaea. And none of them have even a head – they all have only one single **cell**. These microscopic creatures are collectively known as prokaryotes, and they're everywhere on Earth, from the deep sea to our own bodies. Although some bacteria have a bad reputation for causing disease, prokaryotes are important for agriculture, food digestion, and maintaining a healthy planet. We can't live without them!

**WHY IT MATTERS**
By studying prokaryotes, biologists learn how life can exist under extreme conditions or how diseases are spread

**KEY THINKERS**
Antonie van Leeuwenhoek (1632–1723) first saw bacteria under a microscope. Fanny Hesse (1850–1934) discovered that bacteria can grow on agar, which made it easier to study them. Carl Woese (1928–2012) split prokaryotes into two domains, Bacteria and Archaea

**WHAT COMES NEXT**
Because bacteria are easy to grow in a lab, they're used in genetics research to study **DNA** or proteins (see Model organisms, p.20)

In **100** words

The organisms in the taxonomic domains Bacteria and Archaea are collectively known as prokaryotes. They all have just a single cell surrounded by a cell wall. Some prokaryotes have appendages called flagella that help them move around in their environment.
Prokaryotes are everywhere: in soil, in water, and on and inside of larger organisms. The human gut is host to trillions of bacteria that help digest our food. Some archaea can live in extreme conditions, such as in hot springs.
Many useful processes involve prokaryotes, such as food production, digestion or nutrient cycles. However, some bacteria can cause infectious diseases.

## In 100 words

Eukaryotes belong to the domain Eukarya. These **organisms** have more advanced cells than prokaryotes, with cell nuclei and specialised compartments. Eukaryote cells were formed around 2.5 billion years ago by a merging of prokaryote cells. Eukaryotes cover a wide variety of life, including all fungi, plants, animals, and any other organism with a cell nucleus. With the exception of some fungi and most members of the protist kingdom, most eukaryotes have multiple cells that form tissues and organs. Eukaryotes vary in size from tiny, **single-celled** algae to enormous whales and trees. Most of the **biomass** on Earth consists of eukaryotes.

# Eukaryotes

Unlike prokaryotes, eukaryotes have cells with a **nucleus** and other specialised compartments. There is a wide variety of organisms within this domain. They can be as small as a single **cell**, or as big as a redwood tree. Eukaryotes include all plants and animals, as well as fungi and protists such as algae.

**WHY IT MATTERS**
Because all eukaryote cells have things in common, studying one eukaryote organism provides information about all of them (see Cells, p.26 and Model organisms, p.20)

**KEY THINKERS**
Roger Stanier (1916–1982) distinguished eukaryotes from prokaryotes.
Lynn Margulis (1938–2011) showed how eukaryote cells acquired **organelles** from symbiosis of prokaryotes

**WHAT COMES NEXT**
Fossils from extinct eukaryotes and information from eukaryote genes help us to understand the way that different species have evolved (see Evolution, p.18 and Vertebrates, p.17)

**SEE ALSO**
Prokaryotes, opposite
Cells, p.26
Symbiosis, p.70

# 4

# Plants

Plants aren't just for decoration – they're necessary for many other life forms. They produce the oxygen we breathe and the food we eat. We turn plants into medicine, fuels, homes, furniture, books and much more.

Most plants are flowering plants that produce seeds, but there are also plants that reproduce in a different way. For example, ferns use spores.

**WHY IT MATTERS**
Plants help to keep ecosystems in balance. They produce oxygen through photosynthesis and are a source of nutrients for many species (see Photosynthesis, p.58)

**KEY THINKERS**
People in the Middle East, South Asia and the Americas have been cultivating plants for centuries (c. 10,000–8000 BCE). Ji Han (263–307) was a Chinese botanist who systematically documented plants in South East Asia. Nehemiah Grew (1641–1712) described plant parts in detail

**WHAT COMES NEXT**
Learning how to cultivate plants was key to the development of agriculture (see Agriculture, p.162)

**SEE ALSO**
Photosynthesis, p.58
Biodiversity, p.63
Food webs and biomass, p.66
Drug discovery and development, p.167

In **100** words

Plants use energy from the sun to create their own nutrients. They release oxygen to the atmosphere, and this makes them indispensable to many other **species**. Most plants reproduce by sharing pollen from one flower to another, often with the help of insects, after which the flower produces seeds. Humans have learned how to breed plants as a reliable source of food, but plants have also been used as medicine for thousands of years.
The systematic study of plants is called botany. Modern botanists often study plant **DNA** to gather information about different plant species and to help preserve biodiversity.

## In 100 words

The eukaryote kingdom Fungi includes many different organisms, from **single-celled** yeast to enormous mushrooms. All fungi have in common that their **cell** wall includes a molecule called chitin. Most fungi reproduce via spores. Fungi are often found in the soil or on plants, where they play an important ecological role by decomposing old organic matter and returning nutrients to the **ecosystem**. Other fungi are used in food and drink fermentation: for example, to make bread, beer or wine.

While some yeast and mould can cause fungal infections in humans, other fungi can be a source of antibiotics such as penicillin.

# Fungi

If you take a walk through the woods, you're surrounded by fungi. They include the mushrooms growing on the forest floor as well as much smaller **organisms** living in the soil or on tree roots. Fungi play a part in **decomposing** dead plants and bringing **nutrients** back into the soil. Some fungi even made their way into our kitchen as edible mushrooms or baker's yeast, while others are important in medical research – either as medicine or as **pathogens**. Fungi may be low profile compared to animals and plants, but they make a big impact.

**WHY IT MATTERS**
Fungi are an important part of nutrient cycles, food production, and drug discovery

**KEY THINKERS**
Ancient Egyptians (c. 1300–1500 BCE) already knew how to use yeast to prepare bread.
Pier Antonio Micheli (1679–1737) discovered that fungi use spores to reproduce.
Elsie Wakefield (1886–1972) described many previously unnoticed species of fungus

**WHAT COMES NEXT**
The brewer's yeast *Saccharomyces cerevisiae* is a model organism used to study genetics and basic eukaryote biology, and to test new drugs (see Model organisms, p.20)

**SEE ALSO**
Nutrient cycles, p.64
Antibiotics, p.165

# 6

# Invertebrates

What do beetles, mollusks, corals, worms and crabs have in common? They're all members of a very large group of invertebrates. This is not a formal **taxonomy** category, but it covers almost all of the creatures in the animal kingdom.

The name "invertebrates" refers to their lack of a backbone. Some invertebrates, such as jellyfish, have no skeleton at all, while others, like insects, have a hard **exoskeleton** on the outside of their body

Invertebrates are an important part of many **ecosystems**. For example, mollusks filter water, and bees act as pollinators.

## WHY IT MATTERS
Invertebrates shape their environment by pollinating plants, filtering water or delivering nutrients (see Agriculture, p.162)

## KEY THINKERS
Insects: Maria Sibylla Merian (1647–1717), William Kirby (1759–1850).
Marine invertebrates: Yoichiro Hirase (1859–1925), Shintaro Hirase (1884–1939), William Stimpson (1832–1872).
Various invertebrates: Libbie Hyman (1888–1969)

## WHAT COMES NEXT
Research on invertebrates such as the fruit fly *Drosophila melanogaster* and the roundworm *Caenorhabditis elegans* show how genetic information is regulated in all organisms (see Gene expression, p.35 and Model organisms, p.20)

## SEE ALSO
Biodiversity, p.63
Coral reefs, p.72

## In
# 100
## words

Invertebrates include a wide variety of **organisms** in most phyla of the animal kingdom. More than 95 per cent of animal **species** are invertebrates, and new ones are still regularly discovered.
The largest invertebrate is the giant squid, which can grow to more than 10 metres long. Most invertebrates are much smaller; for example, insects, spiders, worms, and smaller marine invertebrates. Invertebrates live on land and in water. Several can fly. Many play important roles in maintaining their **habitats**; for example, insects transfer pollen between flowers to help plants reproduce, and clams filter phytoplankton from water to keep it clear.

Vertebrates are all members of the phylum Chordata in the kingdom Animalia. Humans are vertebrates, as are other mammals, fishes, birds, reptiles, amphibians, and several extinct **organisms** such as dinosaurs.
They live on land, in water, or both. The majority of vertebrates lay eggs, except for most mammals. Vertebrates vary in size from a fly-sized frog to the enormous blue whale, which is the length of six cars.
All vertebrates have jaws and limbs, as well as a similar body pattern and circulation. Recognising similarities between vertebrates has helped biologists to understand evolution and to learn more about human health.

# Vertebrates

Fish, birds, reptiles, amphibians, and mammals might seem like very different creatures, living in wildly varying **habitats**, but they have one thing in common: they all have a backbone, and that makes them vertebrates.

Vertebrates make up less than 5 per cent of all animal **species** and take up less than a full phylum in the animal kingdom. They just happen to be the animals we're most familiar with and the group that includes ourselves.

Studying other vertebrates helps us better understand human biology and evolution.

**WHY IT MATTERS**
Vertebrates include humans as well as most common domestic animals and farm animals

**KEY THINKERS**
Aristotle (384–322 BCE) first distinguished vertebrates and invertebrates.
Mary Anning (1799–1847) discovered fossils of prehistoric marine reptiles which proved that species could go extinct.
Tsen-Hwang Shaw (1899–1964) established vertebrate zoology in China and documented many species

**WHAT COMES NEXT**
Fossils of extinct vertebrates have helped palaeontologists and biologists learn more about evolution (see Evolution, p.18)

**SEE ALSO**

# 8

# Evolution

**WHY IT MATTERS**
Evolution explains how
species are related and
where the diversity in
organisms originated

**KEY THINKERS**
Charles Darwin
(1809–1882) described
the theory of evolution
based on natural
selection.
Mary Leakey
(1913–1996) and Louis
Leakey (1903–1972)
showed that the last
common ancestor of
all humans originated
from Africa.
Richard Lenski's
(1956–) ongoing
long-term evolution
studies in bacteria
demonstrate that
spontaneous genetic
variation can lead to
new properties

**WHAT COMES NEXT**
Modern evolutionary
research relies heavily
on comparing the
genetic sequences
of different species
(see Genetic
sequencing, p.156)

Many people throughout history have realised that the similarities between **organisms** aren't a coincidence: we are all related. How we ended up with so many different **species** was one of the biggest questions in biology for a long time. Charles Darwin found an important piece of the puzzle. He suggested that there is natural variation within each species, but that only certain **traits** are passed on to future generations based on pressures from the environment. His most famous example is of finches that adapted to different food sources by evolving different beak shapes.

Darwin's theory of natural selection provided a plausible and testable idea about our own origins and that of all other organisms. It wasn't the only theory of evolution, but it has held up well against more than 150 years of further research.

Our knowledge of evolution is itself evolving. Darwin didn't know that traits were passed on through **genes**, but modern evolutionary biologists do. They may also study other processes such as **epigenetics**, but overall slow **genetic** change through natural selection remains the main contributor to evolution.

> **"Whilst this planet has gone cycling on according to the fixed law of gravity, from so simple a beginning endless forms most beautiful and most wonderful have been, and are being, evolved."**
>
> **Charles Darwin,** naturalist and biologist

Evolution is gradual change over generations. A key driver is selective pressure, from climate, food or predators. Individuals with traits that help them survive are more likely to mate and pass on their genetic information to future generations. New species form when selective pressure creates distinctly different genetics within a **population**. Each branching point in the tree of life has potential to evolve further. For example, birds and dinosaurs are related, even though only one branch survived. All species have a shared common ancestor. Any two species also share a last common ancestor before their evolutionary paths diverged.

In the Galapagos, Darwin found finches with different beak shapes on each island. Their beaks had evolved under selective pressure at each location.

# 9

# Model organisms

**WHY IT MATTERS**
Model organisms can reveal some of the features of genetics, development or neuroscience that all species share

**KEY THINKERS**
Theodor Escherich (1857–1911) discovered the *bacteria Escherichia coli*. Abbie Lathrop (1868–1918) bred model organism mice. Sydney Brenner (1927–2019) first used the *Caenorhabditis elegans* worm to study neuronal development. Christiane Nüsslein-Volhard (1942 –) studied development in fruit flies

**WHAT COMES NEXT**
**Cell** culture or computational analysis can replace or reduce the use of animals in some types of experiments (see Cell culture, p.143, and Bioinformatics, p.158)

**SEE ALSO**
Gene expression, p.35
Embryonic development, p.88

You won't find model **organisms** on the runway, but they do get a lot of attention. Biologists study model organisms to understand processes that many other **species** have in common.

Because they're widely used, we know a lot about model organisms such as their entire genetic **code**.

Model organisms are usually easy to keep and take care of and have attributes that make them easy to study. For example, zebrafish embryos are transparent, which makes them useful models for vertebrate development. The fruit fly *Drosophila melanogaster* is used in genetics because it breeds quickly and clearly shows the effect of genetic changes. The bacteria Escherichia coli and the yeast *Saccharomyces cerevisiae* are very easy to grow in a lab and widely used in genetics, microbiology or biochemistry research.

Small mammals such as mice are genetically very similar to those of humans, which is why biologists study them as a model to understand mechanisms of disease. In many countries, ethics boards oversee animal research to make sure that researchers consider alternatives where available and treat animals humanely when they work with them.

**In**
**100**
**words**

Model organisms are species that many biologists study to learn about genetics, development or neuroscience. These organisms are models for other species like them. For example, zebrafish and mice are models of vertebrates, while flies and yeast are studied to show general concepts of eukaryote genetics. This is possible because all species are related through evolution.

Model organisms have been kept in labs for many generations and their genetic code is often well understood. Generalising from one species to another doesn't always work, so biologists might study similar processes in a few different species to be certain of their findings.

"I immediately loved working with flies. They fascinated me, and followed me around in my dreams."
**Christiane Nüsslein-Volhard,** biologist

# 10

# Viruses

**WHY IT MATTERS**
Many viruses affect
the health of humans,
animals or plants.
Viral epidemics or
pandemics can have
large-scale effects
on society

**KEY THINKERS**
Martinus Beijerinck
(1851–1931) discovered
the first viral
infection in plants.
June Almeida
(1930–2007)
first identified
coronaviruses.
Françoise Barré-
Sinoussi (1947–) and
Luc Montagnier
(1932–2022) showed
that HIV causes AIDS

**WHAT COMES NEXT**
The development of
vaccines made it
possible to protect
against viruses (see
Vaccines, p.166)

**SEE ALSO**
Gene expression, p.35

For something so small, viruses sure cause a lot of problems. They're not technically considered alive because they need to infect an **organism**'s cells to produce more copies of their **genetic** material. But in doing so, viruses cause a range of diseases in plants, animals and humans.

There are thousands of different viruses, and they can rapidly evolve into new variants. Not all viruses are damaging to humans, but even viruses that affect plants or animals can severely disrupt agriculture or **ecosystems**. Viruses can be grouped into different families and subfamilies based on whether their genetic material is **DNA** or **RNA**, what type of **protein** coat (capsid) they have, and whether or not they have an additional outer layer called an envelope. For example, adenoviruses have DNA inside an icosahedral capsid without an envelope, while coronaviruses have RNA in a helical (twisted) capsid structure that's enclosed in a lipid envelope with proteins.

Virus families and
subfamilies have
different capsid
shapes and
envelopes.

Viruses are tiny microbes that infect the cells of living organisms. They have their own genetic material, surrounded by a protective protein coat, but they need a living cell's machinery to make more copies of themselves. The host **cell** often dies in the process.

There are many different types of viruses. Several are linked to disease in humans, ranging from rhinoviruses causing the common cold to the human immunodeficiency virus (HIV) that causes AIDS.

Since viruses don't have cells, they're not considered to be living things, but they are close enough that the study of viruses is part of biology.

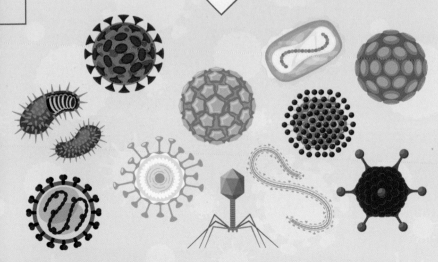

23

# Building Blocks

It might not seem as if flies, yeast or bacteria have much in common with humans. But once you zoom in and look closely, we're actually very similar.

All living things share some common building blocks. They all have at least one or more cells, and DNA with genetic code to form proteins. They all have lipids and carbohydrates and have found ways to store or generate energy.

Because many building blocks of life are so similar between organisms, we can learn a lot about ourselves from studying flies, yeast or bacteria.

# 11

# Cells

**WHY IT MATTERS**
Cells form tissues
and organs, so
understanding
individual cells is a
step to understanding
entire organisms

**KEY THINKERS**
Robert Hooke
(1635–1703) introduced
the term "cell".
Rudolf Virchow
(1821–1902)
Christian de Duve
(1917–2013)
Albert Claude
(1899–1983)
George Palade
(1912–2008)
Jennifer Lippincott-
Schwartz (1952–)

**WHAT COMES NEXT**
Because cells are
only visible with a
microscope, nobody
knew about them until
the 17th century. With
advances in microscope
technology, biologists
are getting more detail
about the way that cells
function (see Electron
microscope, p.144)

**SEE ALSO**
Cell signalling and
transport, p.50
Tissues and
organs, p.106

Cells are a lot like kitchens. Just as every space in a kitchen has a purpose — the oven for cooking, the refrigerator for cooling — so the different compartments in a cell each have their own function.

In eukaryote cells, these compartments include a **nucleus** for storing and transcribing **genetic** material, endoplasmic reticulum where **proteins** are formed, mitochondria that produce energy, Golgi apparatus and vesicles that move content around the cell, and lysosomes that break things down.

Each cell is surrounded by a cell **membrane** made of lipids and proteins. The **organelles** inside the cell have membranes too, to separate their contents from the rest of the cell.

Plant cells are slightly different from animal cells, as they have a hard cell wall surrounding them and contain chloroplasts that help them carry out **photosynthesis**.

All living **organisms** are made of cells. Some have only a single cell, while others have many different types of cells that each have a specific function. These cells group together to form **tissues** and organs.

Within each eukaryote cell are specialized compartments called organelles. Each organelle also has a unique role. For example, the nucleus stores **DNA**, and the mitochondrion produces energy. Each cell and the organelles within it are surrounded by a flexible membrane made of lipids and proteins.

Cells can divide and multiply – for example, during embryonic **development** and growth, or to replace and renew old cells.

Animal cell with organelles that each have a specialised role.

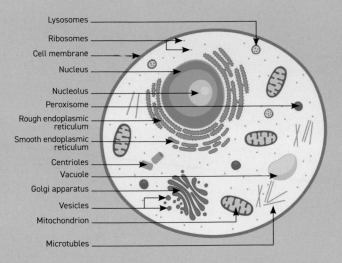

Lysosomes

Ribosomes

Cell membrane

Nucleus

Nucleolus

Peroxisome

Rough endoplasmic reticulum

Smooth endoplasmic reticulum

Centrioles

Vacuole

Golgi apparatus

Vesicles

Mitochondrion

Microtubles

# Nucleus

**WHY IT MATTERS**
The nucleus stores and manages most of the cell's DNA, which in turn contains the genetic information for the entire organism

Let's step into the office, or, as it's called in the **cell**, the nucleus. The nucleus of a cell is an important **organelle** because it acts as a control centre that decides which of the many **genes** in the **genome** need to be active in that cell.

It also includes everything the cell needs to ensure that its **DNA** is of good quality and to copy **genetic** information from DNA to **mRNA**.

A nuclear **membrane** separates the nucleus from the rest of the cell, which is known as the **cytoplasm**.

**KEY THINKERS**
Robert Brown (1773–1858) coined the term nucleus. Albrecht Kossel (1853–1927) identified nucleic acids in the cell nucleus. Jean Brachet (1909–1988) showed that DNA was stored in the nucleus. Marie Daly (1921–2003) developed methods to study the cell nucleus

**WHAT COMES NEXT**
Transcribing DNA to mRNA in the nucleus is the first step of gene expression, which turns genetic information into amino acids and proteins (see Gene expression, p.35)

"Here we have an organ of the cell whose structure and function must be associated with the general processes of life."
**Albrecht Kossel,** biochemist

The nucleus is the largest organelle in a eukaryote cell. It stores most of the cell's DNA, and transcribes genetic information from DNA to messenger **RNA** (mRNA), which then leaves the nucleus through pores in the nuclear membrane, ready to translate to **proteins** in the cytoplasm. Not all of the nucleus is devoted to DNA. The nucleolus, an area within the nucleus, produces ribosomal RNA (rRNA), which forms part of the ribosomes that translate mRNA to proteins in the cytoplasm of the cell.

The nucleus also contains proteins to protect, transcribe and carry out quality control of the cell's DNA.

# 13

# DNA

## WHY IT MATTERS
DNA contains the blueprint for cells to produce what the organism needs to grow and live

## KEY THINKERS
Albrecht Kossel (1853–1927) identified the bases in DNA and RNA. Martha Chase (1927–2003) and Alfred Hershey (1908–1997) showed that DNA holds genetic information. Francis Crick (1916–2004), James Watson (1928–), Rosalind Franklin (1920–1958) and Maurice Wilkins (1916–2004) discovered that DNA is a double-stranded helix

## WHAT COMES NEXT
Genes are parts of DNA that include code to make proteins (see Genes, p.34)

## SEE ALSO
Chromosomes, p.32
Gene expression, p.35
Inheritance, p.86

Even though **DNA** contains all the **genetic** information to make an entire complex **organism**, it's actually a relatively simple molecule. It's big, but it has lots of repeating parts. Both DNA (deoxyribonucleic acid) and **RNA** (ribonucleic acid) are nucleic acids, but there are a few small differences between the two. For example, RNA has the sugar ribose in its backbone, while DNA has deoxyribose with one fewer oxygen **atom**.

DNA is a double helix, with two **strands** that are connected by pairing **bases** together in a consistent pattern so that both strands of DNA include the same information.

Because the only variable parts in a DNA molecule are the order in which the four bases appear, it's possible to describe a DNA **sequence** just by its base letter abbreviations — for example, "ATGACTGTCAGG".

> "DNA is like a computer program but far, far more advanced than any software ever created."
>
> **Bill Gates,** former CEO of Microsoft

DNA forms a helix with two strands.
Each strand has a backbone of alternating
sugar and phosphate molecules. Each sugar group
is connected to one of four different bases: adenine (A),
cytosine (C), guanine (G) or thymine (T). One combined unit
of a sugar, phosphate, and base is a **nucleotide**. The two
strands are connected by hydrogen bonds between
the bases, which always pair A with T and C with
G. RNA molecules are slightly different: they're
usually a single strand, have a uracil
base instead of thymine, and
their sugar group has an
oxygen atom that
DNA doesn't
have.

In the DNA double helix,
thymine is paired with
adenine, and guanine is
paired with cytosine.

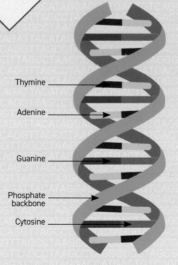

Thymine

Adenine

Guanine

Phosphate
backbone

Cytosine

31

14

# Chromosomes

**WHY IT MATTERS**
Getting long strands
of DNA packed into
a cell nucleus and
making sure genetic
information is passed
on to offspring is
no easy feat, but
folding DNA up into
chromosomes makes
it possible

**KEY THINKERS**
Theodor Boveri
(1862–1915)
Nettie Stevens
(1861–1912)
Barbara McClintock
(1902–1992)
Elizabeth Blackburn
(1948–)
Carol Greider (1961–)

**WHAT COMES NEXT**
Splitting chromosome
pairs between gametes
is the first step in
genetic inheritance
(See Inheritance, p.86)

**SEE ALSO**
Mitosis and
meiosis, p.80
Ageing, p.102

If you stretched out all the **DNA** from just one of your **cells**, it would be about as tall as you are, or even longer. To fit all of that into a cell **nucleus**, it has to be tightly folded. This folded-up shape is called the chromosome. Only a small part of the chromosome is uncoiled at any given time to transcribe **genes**. The rest stays folded to protect it.

All chromosomes come in pairs. Humans have 23 pairs (46 chromosomes in total), but other **species** may have different numbers. Cats have 19 pairs, for example.

When **gametes** (sperm and egg cells) are formed, they each get one half of each pair. Only the sex chromosomes X and Y can form a non-matching pair. In mammals, a biologically female individual has two X chromosomes, while a biologically male individual has X and Y (so a sperm cell either has an X or a Y). In other **organisms** this can be different. For example, female birds have two different sex chromosomes while male birds have a matching pair.

Inside the cell nucleus, DNA combines with **proteins** to form chromatin fibres that are tightly wound into chromosomes, which are visible under a microscope. Chromosome tips are protected by sections of DNA called **telomeres**.

Almost all human cells have 23 pairs of chromosomes, including a pair of sex chromosomes (either XX or XY). Gametes (sperm and egg cells) pass on one copy of each chromosome.

A **sequence** of DNA in a matching location on a chromosome pair is called an **allele**. If both alleles are the same, this **genetic** sequence is homozygous in that individual. If it's different, it's heterozygous.

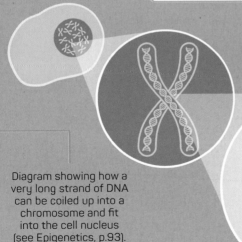

Diagram showing how a very long strand of DNA can be coiled up into a chromosome and fit into the cell nucleus (see Epigenetics, p.93).

# 15

# Genes

When biologists first started thinking about heritable traits, they used the term *gene* to talk about the thing that was inherited. It took several decades to figure out that a gene is a segment of **DNA**.

Humans have about 20,000 genes, but the numbers are different in other **organisms**. The fruit fly has about 16,000, for example, but a potato has 39,000.

Genes take up only a small amount of our DNA. The rest of it is non-coding, but some of this non-coding DNA still has an important role.

**WHY IT MATTERS**
Genes include the instructions to regulate all functional processes in an organism. They're the carriers for heritable traits passed on between generations

**KEY THINKERS**
Wilhelm Johannsen (1857–1927) coined the term *gene*.
George Beadle (1903–1989) and Edward Tatum (1909–1975) showed that each gene encodes one protein.
Esther Lederberg (1922–2006) showed how bacterial genes are regulated

**WHAT COMES NEXT**
Because they are instructions, genes don't play an active role until these instructions are followed by transcribing genes into **mRNA** which is translated to proteins (see Gene expression, opposite)

**SEE ALSO**
Transcription, p.36
Translation, p.38
Genetic sequencing, p.156

## In 100 words

A gene is a piece of DNA that includes the **code** to produce **RNA** or **proteins**, which in turn carry out a function.
Genes take up only a small part of DNA. An estimated 1–2 per cent of human DNA is coding DNA, covering about 20 thousand genes. Other organisms have different numbers of genes.
In between genes is non-coding DNA. This includes gene regulators as well as DNA needed for structure (for example, **telomeres** on chromosomes), but a lot of non-coding DNA is not yet fully understood, and some of it may have no clear purpose at all.

In
**100**
words

Gene expression turns a gene into a functional product. For most genes, that functional product is a protein.

Even though all cells have the entire organism's DNA in their chromosomes, not every cell expresses every gene all the time. Gene expression is regulated so that each gene only makes RNA and proteins in the cells where they're needed at that moment. Specialised cells, such as **neurons**, will express the genes required for that type of cell. In eukaryotes, gene expression starts with transcription of **DNA** to **mRNA** in the **nucleus**, followed by translation of mRNA to protein in the **cytoplasm**.

## WHY IT MATTERS
Gene expression is the process that converts genotype to phenotype. Every trait in an organism is determined not just by the genetic code itself, but by how that code is expressed

## KEY THINKERS
Barbara McClintock (1902–1992)
Francis Crick (1916–2004)

## WHAT COMES NEXT
The first step of gene expression is to copy the DNA code from the gene to mRNA (see Transcription, p.36)

## SEE ALSO
Epigenetics, p.92
Homeostasis, p.130
Manipulating gene expression, p.152

# Gene expression

**Genes** include the entire blueprint for an **organism**, from physical appearance to how all the organs function. How can a simple DNA **strand** have such big consequences? It's all possible through gene expression.

In gene expression, **genetic** information is copied from DNA to **RNA** and then to **proteins**. Proteins (and some RNA) carry out all kinds of processes within the body, from "creating pigment for hair and eyes" to "keeping the heart beating". Gene expression is carefully regulated so that every **cell** activates the right genes at the right time for that cell's purpose.

# 17

# Transcription

**WHY IT MATTERS**
DNA is stuck in the nucleus in the form of chromosomes, but all the machinery to make proteins is in the cytoplasm. Transcribing DNA to mRNA creates a copy of the code that can be transported to the site of protein production

**KEY THINKERS**
François Jacob (1920–2013)
Jacques Monod (1910–1976)
Roger Kornberg (1947–)

**WHAT COMES NEXT**
After transcription, mRNA is translated to proteins (see Translation, p.38)

**SEE ALSO**
Nucleus, p.28
Enzymes, p.48
Epigenetics, p.92

Before **DNA** becomes **protein**, there's an intermediate step. An enzyme called **RNA polymerase** copies (or transcribes) **genetic** information from DNA to messenger RNA (**mRNA**). This is such an important process that all living **organisms** and even some viruses have an RNA polymerase.

Not every **gene** needs to be active all the time, so the RNA polymerase needs to be guided to the right place. Part of this process is done by transcription factors which bind to non-coding regions close to the start of the gene if that gene needs to be turned on or off.

The area of DNA where the transcription factors and RNA polymerase bind is called the promoter site of a gene. This is where RNA polymerase starts reading DNA and creating a copy of the message to mRNA. Depending on the gene, the initial transcript might still have some sections that aren't needed for the final protein. These "introns" are removed before the mRNA moves to the **nucleus** to be turned into protein.

Introns are regions within some genes that are removed at the mRNA stage. The exons remain and include the code for a protein.

The first step of gene expression is the transcription of DNA to mRNA by the enzyme RNA polymerase. For transcription to take place, the gene must have another protein – a transcription factor – bound to a promoter site in the non-coding region just before the start of the gene. Transcription factors regulate gene expression by determining which genes interact with RNA polymerase.

Eukaryote mRNA is further processed to splice out introns (non-coding sections within some genes) before the final mRNA is transported to the **cytoplasm** where it will be turned into proteins. This way, DNA itself can stay in the nucleus.

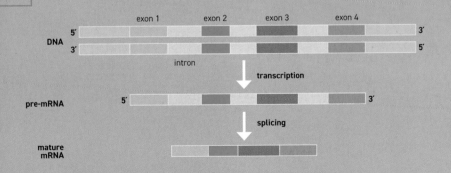

37

# 18

# Translation

After transcription, the next step of gene expression is translation from **mRNA** to **proteins**. This is carried out by ribosomes.

A ribosome is a complex of proteins and ribosomal **RNA** (rRNA) found in every living **cell**. It recognises the start of an mRNA molecule and starts reading the **code**.

Every group of three nucleotides is a "**codon**" that recruits a specific transfer RNA (tRNA) molecule to attach an **amino acid** to the growing protein chain. The first codon is always the start codon AUG, which starts the protein chain with a methionine molecule. Additional amino acids are connected to the previous ones by peptide bonds to create a chain.

The four nucleotides in RNA combine into 64 unique codons, which is more than enough for all human amino acids as well as for signals to start and stop the chain. Some of them are duplicates, which means that sometimes a tiny variation in the **gene** can still produce the same protein. For example, both CGC and CGA are code to add the amino acid arginine to the protein in that spot. When the ribosome reaches a "stop" codon, it lets go of the protein and mRNA.

In
**100**
words

After gene transcription, mRNA molecules
are translated into proteins in the **cytoplasm**.
Every three nucleotides of mRNA form a codon
that codes for a specific amino acid or a "start" or "stop"
signal. When mRNA is passed through a ribosome complex,
each codon is matched to an anti-codon on a transfer
RNA (tRNA) molecule that is connected to an amino acid.
The amino acids are connected in a chain
to form a protein.
Even though hundreds of amino acids
exist in nature, only 20 types of
amino acids are used to
make all of the proteins
in the human
body.

Every group of three nucleotides in the mRNA transcript is a codon that either matches a unique amino acid or instructs the ribosome to stop translating. The "start" signal is always a methionine amino acid.

Second Letter

| First Letter | | U | C | A | G | Third Letter |
|---|---|---|---|---|---|---|
| **U** | | UUU UUC }Phe  UUA UUG }Leu | UCU UCC UCA UCG }Ser | UAU UAC }Tyr  UAA **Stop** UAG **Stop** | UGU UGC }Cys  UUU **Stop** UUC Trp | U C A G |
| **C** | | CUU CUC CUA CUG }Leu | CCU CCC CCA CCG }Pro | CAU CAC }His  CAA CAG }Gln | CGU CGC CGA CGG }Arg | U C A G |
| **A** | | AUU AUC AUA }Ile  AUG Met | ACU ACC ACA ACG }Thr | AAU AAC }Asn  AAA AAG }Lys | AGU AGC }Ser  AGA AGG }Arg | U C A G |
| **G** | | GUU GUC GUA GUG }Val | GCU GCC GCA GCG }Ala | GAU GAC }Asp  GAA GAG }Glu | GGU GGC GGA GGG }Gly | U C A G |

# 19

# Protein structure

**WHY IT MATTERS**
Studying the structure of proteins often gives important clues about its function. Some conditions, such as Alzheimer's disease, are affected by misfolded proteins

**KEY THINKERS**
Linus Pauling (1901–1994)
Dorothy Hodgkin (1910–1994)
John Kendrew (1917–1997)
Max Perutz (1914–2002)

**WHAT COMES NEXT**
Proteins are too small to see under a regular microscope, but their structure can be understood via X-ray crystallography (see X-ray crystallography, p.146)

**SEE ALSO**
Mutation and variation, p.42
Enzymes, p.48

Thousands of different **proteins** all have their own role in keeping you and other living **organisms** alive. To name just a few: **haemoglobin** carries oxygen, **myosin** makes muscles move, and insulin signals that sugar needs to be processed. They all have unique tasks and can only carry out their task if they are folded into the right shape.

Even though the **amino acid** sequence can be predicted from the **DNA code**, the final three-dimensional structure of a protein is more complex. Parts of the protein that were far apart in the amino acid sequence can end up close together in its final folded form.

The primary structure of a protein is its amino acid sequence. Secondary structures are patterns such as helices or sheets that form a mini-structure within the protein. The tertiary structure is the entire folded protein. Some proteins also have a quaternary structure by forming a complex with other proteins.

The end result is that proteins have some rigid parts and some flexible parts that together form a coherent shape that lets it interact with other molecules to carry out its role.

The function of a protein is not only determined by its amino acid sequence (primary structure) but also by the way the chain is folded into local structures (secondary structures) and how the overall three dimensional protein looks (tertiary and quaternary structures).

## In 100 words

The primary structure of a protein is its amino acid sequence. The amino acids interact with each other and with their environment to form a three-dimensional structure.

Some amino acids are small, while others take up a lot of space. Some amino acids prefer being on the outside of a protein, others on the inside. Folding patterns include secondary protein structures such as α-helixes or β-sheets. These and other folds determine the eventual tertiary or three-dimensional structure of the protein.

Many proteins interact with other molecules, so certain amino acids in the structure need to be available for that interaction.

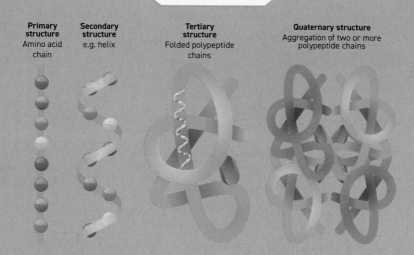

**Primary structure**
Amino acid chain

**Secondary structure**
e.g. helix

**Tertiary structure**
Folded polypeptide chains

**Quaternary structure**
Aggregation of two or more polypeptide chains

# 20

# Mutation and variation

Genetic variation shows up as differences between individuals. It's why not everyone likes coriander or has the same earlobes. These mutations can be passed on to the next generation.

Some mutations cause a **protein** to stop working, which can lead to disease or even stop an **organism** from fully developing. Most **genes** have two copies (one on each chromosome), so some of these effects are noticeable only if a **genetic variant** is inherited from both parents.

Mutations have a few different effects:
· Insertion: adding a **nucleotide**
· Deletion: removing a nucleotide
· Substitution: a different nucleotide takes its place

Inserting or deleting three nucleotides in a row changes only one **amino acid** in the protein. But inserting or deleting one single nucleotide changes the reading frame of the **mRNA**, resulting in a very different protein.

Substituting one nucleotide may result in a different amino acid, which can be harmless depending on how it affects the overall protein structure and function.

Over time, genetic mutation and variation that does not harm the organism can lead to the gradual **development** of different **species**.

There are minor genetic differences in the **DNA** of individuals of the same species. These small variations can appear over time through genetic mutations. Many mutations are harmless and lead to diversity within species. Mutations are inherited only if they occur in egg or sperm cells. Genetic mutations can completely change the structure and function of a protein. Even small variations in DNA affect which amino acids are added during translation, which changes the resulting protein. If that change is disruptive, it can stop the protein from functioning properly. Several diseases are linked to genetic mutations that affect protein function.

Example to show how inserting, deleting or substituting a nucleotide in the gene can affect the amino acids in the protein. (See also the codon table on p.39)

| | Example DNA Sequence | Corresponding mRNA codons | Amino Acids |
|---|---|---|---|
| Original | ATGCATTACCCG | AUG CAU UAC CCG | Met His Tyr Pro |
| Insertion Example | ATGCATTAACCCG | AUG CAU UAA CCC | Met His STOP (end of chain, rest is not translate |
| Deletion Example | ATGCTTACCCG | AUG CUU ACC | Met Leu Thr |
| Substitution Example | ATGGATTACCCG | AUG GAU UAC CCG | Met Asp Tyr Pro |

# 21

# Carbohydrates

Sugars, fibres and starches are all examples of carbohydrates. These molecules play an important role in all living **organisms** by providing structural support and storing energy. The most effective way for humans to get carbohydrates is through starchy food such as bread or rice.

Carbohydrates are polymers, just like **proteins** and lipids. This means they're made of multiple connected subunits or monomers. In proteins, the monomers are **amino acids**. In carbohydrates, they're monosaccharides.

Some examples of monosaccharides are glucose, fructose, galactose and ribose. Carbohydrates can include different types of monosaccharides, and the chains can form branches.

**WHY IT MATTERS**
Carbohydrates store energy and form structural components such as cell walls

**KEY THINKERS**
Claude Bernard
(1813–1878)
Gerty Cori (1896–1957)
Carl Cori (1896–1984)

**WHAT COMES NEXT**
Cells can release energy from carbohydrates by processing glucose (see Glycolysis, p.54)

**SEE ALSO**
Photosynthesis, p.58
Nutrition and digestion, p.126

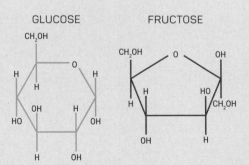

Glucose and fructose are two examples of monosaccharides.

GLUCOSE

FRUCTOSE

Carbohydrates store energy in living organisms and are part of structural components, such as the **cell** wall of plant cells.

They are polymers, which are large molecules made of several subunits, in this case monosaccharides. Carbohydrates can combine different types of monosaccharides and form branches, so there is a lot of variety in size or shape. Monosaccharides are simple sugars such as glucose, fructose or ribose (which forms part of **RNA** and **DNA**). Disaccharides combine two monosaccharides. For example, sucrose is formed of glucose and fructose. More complex carbohydrates such as starch or cellulose form even longer chains called polysaccharides.

Monosaccharides can combine to form disaccharides or larger polysaccharides.

Monosaccharide

Disaccharide

Polysaccharide

# 22

# Lipids

**WHY IT MATTERS**
Lipids are one of
the main building
. blocks of cells in all
living organisms

**KEY THINKERS**
Components of lipids:
Michel Chevreul
(1786–1889)
Phospholipids:
Théodore Gobley
(1811–1876)
Cell membranes:
Charles Ernest
Overton (1865–1933)

**WHAT COMES NEXT**
Phospholipid
membranes play an
important role in
transporting cargo
within and between
cells (see Cell
signalling and
transport, p.50)

**SEE ALSO**
Cells, p.26
Citric acid cycle, p.56
Nutrition and
digestion, p.126

Lipids is the collective name for oils, fats, waxes and other molecules that don't dissolve in water. They mostly consist of carbon and hydrogen **atoms**. In human biology, lipids are important for energy storage and for the formation of cellular membranes.

Many of the most common lipids in plants and animals are triglycerides. These are made of a glycerol molecule with three fatty acids attached to it. Triglycerides store energy that is released when their fatty acid chains are broken down.

Fatty acids have long chains of carbon and hydrogen atoms. They can be either saturated (with hydrogen atoms) or unsaturated. Unsaturated fatty acids have a double bond between two carbon atoms that slightly changes the shape of the fatty acid chain.

Phospholipids are similar to triglycerides, but with a phosphate molecule instead of one of the fatty acid chains. This gives them a **hydrophilic** head and a **hydrophobic** tail, which allows phospholipids to form a double-layered **cell membrane**.

Mammals can't produce all types of lipids on their own, so they need to get some fats through their diet.

# In 100 words

Lipids include a wide group of organic molecules with the common property that they're not soluble in water (hydrophobic). Important types of lipids in biology include triglycerides, which store energy in the form of fat, and phospholipids, which form membranes around cells and **organelles**. Triglycerides have three fatty acid chains connected to a glycerol molecule. In phospholipids, one of those fatty acid chains is replaced with a phosphate group, which gives the phospholipid a water-soluble (hydrophilic) head and a hydrophobic tail. Fatty acid chains can vary in length. Unsaturated chains have double bonds between carbon atoms which change their shape.

BELOW LEFT A phospholipid has two fatty acid chains that form a hydrophobic tail. The hydrophilic head includes phosphate and glycerol.

BELOW Phospholipids can combine to form a phospholipid bilayer – in cell membranes, for example.

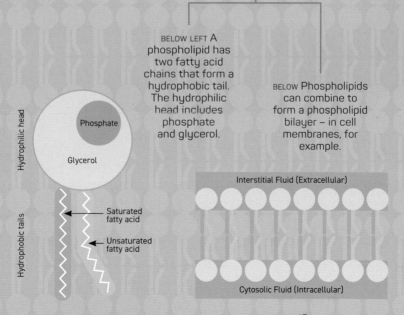

Hydrophilic head

Hydrophobic tails

Phosphate

Glycerol

Saturated fatty acid

Unsaturated fatty acid

Interstitial Fluid (Extracellular)

Cytosolic Fluid (Intracellular)

# 23

# Enzymes

**WHY IT MATTERS**
Without enzymes,
some of the most
important processes
in living organisms
don't take place

**KEY THINKERS**
Active site: Daniel
Koshland (1920-2007)
Enzyme kinetics: Maud
Menten (1879-1960)
and Leonor Michaelis
(1875-1949)

**WHAT COMES NEXT**
Enzymes play
important roles in
common metabolic
processes such as
photosynthesis and
cellular respiration
(see Citric acid
cycle, p.56 and
Photosynthesis, p.58)

**SEE ALSO**
Protein structure, p.40
Antibiotics, p.165
Drug discovery and
development, p.167

**DNA** transcription, food digestion, energy production and many other biological processes would not happen without **enzymes**.

Many different **proteins** are enzymes, and they all have a specialised task. Despite their differences, enzymes all work in a similar way.

Enzymes convert a substrate (the starting molecule) to a product (the end molecule). The substrate binds to the active site of the enzyme, which includes a binding site as well as a catalytic site that carries out the chemical reaction. The active site can wiggle slightly to give the substrate a perfectly fitting pocket.

Enzyme activity can be regulated by other proteins that act as activators or inhibitors. Some enzymes also get help from other molecules called cofactors or coenzymes. All these interactions change the rate of the reaction catalysed by the enzyme (enzyme kinetics).

Enzyme inhibitors either block the active site or interact with the enzyme in another way to change its function. This happens naturally to regulate enzyme reaction pathways, but inhibitors can also be deliberately designed to act as medication. For example, aspirin blocks an enzyme involved in inflammation, while several types of antibiotics inhibit bacterial enzymes so that the bacteria can't survive.

Enzymes are proteins that convert one molecule (their substrate) to another molecule. They speed up chemical reactions to such an extent that without them a lot of cellular activities simply don't happen. Enzymes have an active site with a binding site and catalytic site. The substrate binds to the binding site while the catalytic site carries out the reaction. The activity of the enzyme and the rate of the enzymatic reaction can be altered by other molecules. These include both activators and inhibitors which can act on different parts of the enzyme. Many drugs work by interfering with specific enzymes.

Enzymes have an active site that interacts with a substrate. This site can slightly adjust shape to fit the substrate. Inside the active site is a catalytic site that converts the substrate to its product.

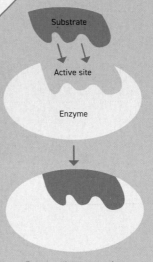

Substrate

Active site

Enzyme

Enzyme-substrate complex

# 24

# Cell signalling and transport

**WHY IT MATTERS**
Proteins and other molecules in a cell need to be in the right location to carry out their function. If they can't physically move in or between cells, they interact with other molecules to pass on information

**KEY THINKERS**
Receptors: John Newport Langley (1852–1925)
Intracellular transport: Marilyn Farquhar (1928–2019)
Ion channels: Erwin Neher (1944–) and Bert Sakmann (1942–)
Vesicles: James Rothman (1950–), Randy Schekman (1948–), and Thomas Südhof (1955–)

**WHAT COMES NEXT**
Cell signalling and transport costs energy, which is accessed in the form of ATP (see ATP, p.52)

**SEE ALSO**
Cells, p.26
Brain and nervous system, p.124

All the components in a **cell** communicate with each other. Some activate or inhibit gene expression; others act as **enzymes** or interact with other molecules to pass on a message.

But **proteins** and other molecules have to be at the right place at the right time to carry out their role. And some molecules will need to pass their message through a cell **membrane** to communicate their signal.

Within the **lipid bilayer** of a cell membrane are proteins that help with cell signalling and transport. Some membrane proteins act as **receptors** that pass on a signal by interacting with different molecules on each side of the membrane. Other membrane proteins form channels that can let some small molecules through.

Anything larger can be encapsulated into a small membrane-bound compartment called a vesicle. These can merge with cell or **organelle** membranes to transport their cargo within or between cells. For example, this is how **neurons** release neurotransmitters to communicate with other cells.

Proteins and other molecules interact with each other in carefully orchestrated signalling pathways. These cell signalling pathways occasionally require a signal to pass a membrane.

Within the lipid bilayer of cell membranes are proteins that can send the signal across. Receptor proteins interact with molecules outside the cell to pass on a signal to other proteins on the inside. Channels can let specific small molecules across a membrane, often actively controlling which molecules pass through and when.

Membranes can also pinch off small lipid vesicles for molecules that need to be moved between cell compartments or outside of the cell.

One way for cells to receive information from their environment is through signal molecules which bind receptors on the cell surface. The receptors then transduce information through a signalling pathway inside the cell.

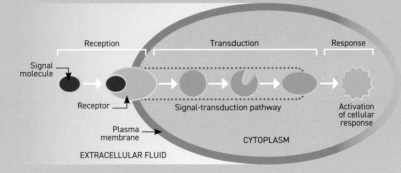

Reception | Transduction | Response

Signal molecule

Receptor

Signal-transduction pathway

Plasma membrane

CYTOPLASM

EXTRACELLULAR FLUID

Activation of cellular response

# 25

# ATP

Nothing happens if you don't apply some energy to it. Within a **cell**, that energy often comes in the form of the molecule adenosine triphosphate (ATP).

ATP can move around the cell to reach places where energy-consuming reactions take place, such as **protein synthesis** or the steps needed for muscle contraction.

ATP releases energy by removing a phosphate group to become adenosine diphosphate (ADP), which can be recycled back into ATP.

**WHY IT MATTERS**
ATP is the energy currency for the cell, making it possible to quickly release energy where it's needed

**KEY THINKERS**
Karl Lohmann (1898–1978) discovered ATP
Fritz Lipmann (1899–1986) proposed ATP as cellular energy source

**WHAT COMES NEXT**
Most ATP is produced either in the mitochondria during aerobic respiration or in chloroplasts during photosynthesis (see Mitochondria and chloroplasts, opposite)

**SEE ALSO**
Glycolysis, p.54
Citric acid cycle, p.56
Photosynthesis, p.58

In **100** words

Within a cell, energy is stored, transported and used as adenosine triphosphate (ATP). ATP is made of an adenine **base**, a ribose sugar, and a tail of three phosphate groups. When one of the phosphate groups is removed by the enzyme ATP hydrolase, energy is released. This energy can be transferred to other molecules to help carry out energy-consuming reactions in the cell. The removed phosphate can also be directly added to other molecules in a phosphorylation reaction. ATP with one phosphate removed is adenosine diphosphate (ADP), which can be turned back into ATP as part of **photosynthesis** and **respiration**.

Mitochondria are present in most eukaryote cells, while chloroplasts are only found in plants and algae. Both organelles play an important role in energy conversion. The main role of mitochondria is to produce the majority of **ATP** that the cell needs. Chloroplasts carry out **photosynthesis** to convert light energy to oxygen and glucose as food for the plant. The pigment chlorophyll in chloroplasts is what gives plants their green colour. Mitochondria and chloroplasts used to be bacteria that were taken up by other cells almost two billion years ago. As a result, they still have some of their own DNA.

# Mitochondria & chloroplasts

Of all the **organelles** in the **cell**, mitochondria and chloroplasts have perhaps the most interesting history. Both of these organelles started out as prokaryotes that were taken up by other cells to become the organelles we know now, with specialised functions in energy conversion.

Mitochondria and chloroplasts have some of their own **DNA**, separate from the DNA in the cell **nucleus**. In humans, mitochondria are inherited only from the mother. So if you include mitochondrial DNA in the overall **genome**, everyone has slightly more maternal DNA than paternal DNA.

**WHY IT MATTERS**
Mitochondria are sometimes referred to as the "powerhouse of the cell" because they produce the majority of ATP. Chloroplasts are responsible for producing oxygen

**KEY THINKERS**
Chloroplasts: Hugo von Mohl (1805–1872)
Mitochondria: Albert von Kölliker (1817–1905)
Evolution of mitochondria and chloroplasts: Lynn Margulis (1938–2011)

**WHAT COMES NEXT**
Both of these organelles are in charge of specialised pathways central to energy production (see Citric acid cycle, p.56, and Photosynthesis, p.58)

**SEE ALSO**
Cells, p.26
Symbiosis, p.70

27

# Glycolysis

**WHY IT MATTERS**
Glycolysis is the first step in turning nutrients such as carbohydrates into energy that cells can use

**KEY THINKERS**
Gustav Embden (1874–1933)
Otto Meyerhof (1884–1951)
Jakub Karol Parnas (1884–1949)

**WHAT COMES NEXT**
Once pyruvate enters the mitochondria, it is turned into acetyl CoA, which enters the citric acid cycle or Krebs cycle, through which most ATP is produced (see Citric acid cycle, p.56)

**SEE ALSO**
Carbohydrates, p.44
Nutrition and digestion, p.126

Running a marathon and brewing beer seem like two very different things, but both start with glycolysis. This is the production of the molecule pyruvate from glucose. It's a two-stage process with several chemical reactions.

Once pyruvate is formed, it can be turned into lactate or ethanol in fermentation processes, such as those that take place in yeast when brewing beer. This is **anaerobic respiration**, which doesn't use oxygen.

Pyruvate can also enter mitochondria, where it's converted to acetyl CoA and enters the citric acid cycle. This process requires oxygen and is called **aerobic** respiration. Aerobic respiration is the main process used when energy needs to last a long time, such as when running a marathon.

# In
# 100
# words

Glycolysis is the first stage of cellular respiration. This is the process of turning **nutrients** into energy. Glycolysis involves several subsequent phosphorylation and oxidation reactions to convert glucose to pyruvate in the **cytoplasm**. These steps create a small amount of **ATP**.
The resulting pyruvate can be converted further in two different processes: aerobic respiration (which requires oxygen) or anaerobic respiration (without oxygen). In anaerobic respiration processes, such as fermentation, pyruvate is converted to ethanol or lactate. In aerobic respiration, pyruvate enters the mitochondria, where it is converted to acetyl coenzyme A (acetyl CoA) which then enters the citric acid cycle.

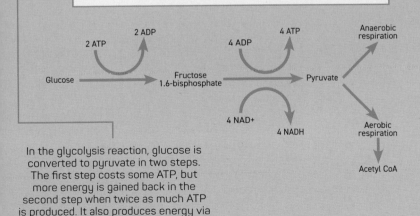

In the glycolysis reaction, glucose is converted to pyruvate in two steps. The first step costs some ATP, but more energy is gained back in the second step when twice as much ATP is produced. It also produces energy via NADH (See Citric acid cycle, p.56).

# 28

# Citric acid cycle

**WHY IT MATTERS**
The citric acid cycle and subsequent oxidative phosphorylation are responsible for the majority of ATP production. They turn nutrients from food into energy your cells can use

**KEY MOMENTS**
Hans Krebs (1900–1981)
Vladimir Engelhardt (1894–1984)
Albert Lehninger (1917–1986)

**WHAT COMES NEXT**
The oxygen needed for aerobic respiration has to come from somewhere. Most animals acquire oxygen through breathing, but only plants and algae can produce it (see Lungs and breathing, p.108, and Photosynthesis, p.58)

**SEE ALSO**
Enzymes, p.48
ATP, p.52
Mitochondria and chloroplasts, p.53
Glycolysis, p.54

The citric acid cycle, or Krebs cycle, is an intricate way of turning food into lots of energy. It starts with acetyl CoA, which can come from carbohydrates via pyruvate but also from fatty acids or **proteins**. Acetyl CoA combines with oxaloacetate to form citrate, also known as citric acid. After several other steps, citrate is gradually turned into oxaloacetate and the cycle starts again.

Along the way, several steps in the cycle produce carbon dioxide ($CO_2$) and **ATP**. The cycle also reduces (adds hydrogen **atoms** to) the coenzymes NAD+ and FAD+. The vast majority of ATP formed from the citric acid cycle is created when these two coenzymes lose their hydrogens again in a process called oxidative phosphorylation. These combined processes are called **aerobic respiration**, because oxygen is used during oxidative phosphorylation.

NAD+ is also reduced in glycolysis and when converting pyruvate to acetyl CoA. Altogether, the full process of aerobic respiration produces 32 ATP for every glucose molecule. And all of this happens in the mitochondria.

## In 100 words

The citric acid cycle is also known as the Krebs cycle. It occurs in the mitochondria and starts with acetyl CoA, which is derived mostly from carbohydrates via glycolysis. Acetyl CoA combines with oxaloacetate to form citric acid, which is gradually converted back into oxaloacetate while releasing ATP and $CO_2$ and reducing the coenzymes NAD+ and FAD+. Reduced NAD+ and FAD+ then produce even more ATP during oxidative phosphorylation. In the process, protons (positively charged hydrogen) are pumped across the inner mitochondrial **membrane**. On their way back, they move through an ATP synthase enzyme in the membrane which produces ATP.

In the citric acid cycle, Acetyl CoA combines with oxaloacetate to become citrate, which produces different forms of energy in several reactions. Eventually, it becomes oxaloacetate again and the cycle starts over.

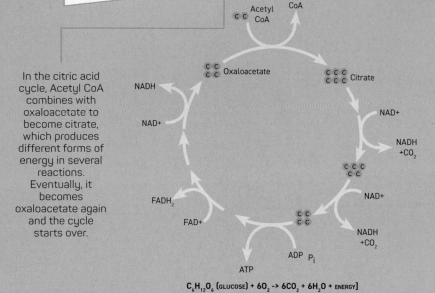

$C_6H_{12}O_6$ (GLUCOSE) + $6O_2$ -> $6CO_2$ + $6H_2O$ + ENERGY]

# 29

# Photosynthesis__

During **photosynthesis**, carbon dioxide and water are converted into glucose and oxygen, using energy from sunlight. In a way, photosynthesis is the opposite of what happens during **aerobic respiration**, where glucose and oxygen are turned into water and carbon dioxide (and energy).

Only plants and algae (and some bacteria) can carry out photosynthesis, so we rely on them to produce oxygen. They're able to do this thanks to the green pigment chlorophyl which is found in the grana – stacks of disc-shaped structures – in the chloroplast.

Chlorophyl absorbs energy from sunlight, which allows it to produce oxygen. This is the light-dependent stage of photosynthesis. During this stage **ATP** is also formed, and the coenzyme NADP is reduced.

The next stage is light-independent and is also called the Calvin cycle. It happens in the liquid-like stroma that fills the main area of the chloroplast. During this process, $CO_2$ is converted to glucose.

The ATP and reduced NADP that were formed in the light-dependent reaction are recycled during the light-independent stage, ready to be used again.

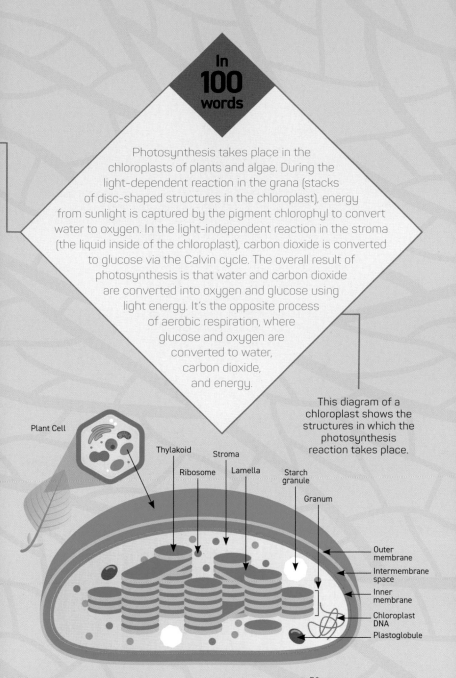

In
**100**
words

Photosynthesis takes place in the chloroplasts of plants and algae. During the light-dependent reaction in the grana (stacks of disc-shaped structures in the chloroplast), energy from sunlight is captured by the pigment chlorophyl to convert water to oxygen. In the light-independent reaction in the stroma (the liquid inside of the chloroplast), carbon dioxide is converted to glucose via the Calvin cycle. The overall result of photosynthesis is that water and carbon dioxide are converted into oxygen and glucose using light energy. It's the opposite process of aerobic respiration, where glucose and oxygen are converted to water, carbon dioxide, and energy.

This diagram of a chloroplast shows the structures in which the photosynthesis reaction takes place.

Plant Cell

Thylakoid

Stroma

Ribosome

Lamella

Starch granule

Granum

Outer membrane

Intermembrane space

Inner membrane

Chloroplast DNA

Plastoglobule

# Ecosystems

Imagine a forest. It has trees, but also flowers, insects, birds, and larger animals like deer. It has soil and rocks and perhaps a lake. All these living and non-living parts interact and together they form the ecosystem of that forest.

There are many ecosystems all over our planet. The lake within the forest is an ecosystem of its own.

Ecosystems can be large or small. Even your intestines are an ecosystem for bacteria! By studying ecosystems, biologists learn how different organisms interact with each other and their environment.

# 30 Populations and ecosystems

Populations are about more than people. They are the total number of any single **species** in a certain area.

Populations interact with other living **organisms** to form communities. When you also add all the non-living components such as soil or weather in that area, you get the complete ecosystem.

Some ecosystems are easier to define than others. Islands and lakes have clear borders, but a national park can have animals moving in and out. Combined with other variables, such as changing **population** sizes, food webs, nutrient cycles, weather and predator–prey relationships, you can see why ecosystems can be difficult to study.

**WHY IT MATTERS**
Studying population changes within the boundaries of an ecosystem reveals how different species are influenced by each other and their environment

**KEY THINKERS**
Population ecology: Eugene Odum (1913–2002)
Ecosystem: Arthur Tansley (1871–1955)
Carrying capacity: Pierre Verhulst (1804–1849)

**WHAT COMES NEXT**
Changes in population levels can indicate a loss of biodiversity within an ecosystem (see Biodiversity, opposite)

**SEE ALSO**
Living in groups, p.68
Symbiosis, p.70

In **100** words

An ecosystem is the combination of all living (biotic) and non-living (abiotic) components within a defined area. All organisms of one species in that area form a population. Different populations within the same ecosystem are a community.
Population size is always fluctuating due to birth, death, migration, or predator–prey relationships.
The carrying capacity is the maximal possible population of a species within a specific ecosystem.
Other aspects of ecosystems also change over time, including food webs or abiotic components such as the weather. All these changes are happening at once, which is what makes ecosystems so complex.

## 31

Biodiversity describes the variety of species within an ecosystem. Populations within an ecosystem depend on each other for nutrients or shelter. Even abiotic components can be affected by loss of a species. For example, loss of trees can lead to erosion after flooding.

It's also important to have **genetic diversity** within species. This ensures that multiple **genetic variants** are present in a **population**, so that a change in environmental conditions in an ecosystem does not put equal pressure on all members of a species. At a larger scale, biodiversity also describes the variety of different ecosystems that exist on Earth.

**WHY IT MATTERS**
Biodiversity keeps everything in balance. A varied and diverse ecosystem is more resilient to sudden changes such as weather extremes or disease outbreaks

**KEY THINKERS**
Thomas Lovejoy (1941–2021)
E.O. Wilson (1929–2021)
Sandra Díaz (1961–)
Brigitte Baptiste (1963–)

**WHAT COMES NEXT**
Conservation is focused on protecting biodiversity between species, while genetic sequencing can show diversity within species (see Conservation, p.163, and Genetic sequencing, p.156)

**SEE ALSO**
Populations and ecosystems, opposite
Mutation and variation, p.42
Nutrient cycles, p.64

# Biodiversity

It would be boring if everyone were the same, but in **ecosystems** it can also be dangerous. Different **organisms** interact with each other and have intertwined lives that are difficult to pry apart. For example, animals need oxygen that's made by plants who get pollinated by insects. Those connections rely on a variety of **species** all playing their part. If one disappears, others may struggle to adjust.

"Every scrap of biological diversity is priceless, to be learned and cherished, and never to be surrendered without a struggle."
**E.O. Wilson,** in *The Diversity of Life*

# 32

# Nutrient cycles

**WHY IT MATTERS**
The constant exchange
of energy and nutrients
keeps life going

**KEY THINKERS**
Ecological balance:
Indigenous people
around the world
(since c. 10,000 BCE)
Nutrients for plant
growth:
Jean-Baptiste
Boussingault
(1802–1887)
Recycling nutrients:
Sergei Winogradsky
(1856–1953)

**WHAT COMES NEXT**
Plants get nutrients
from soil, and animals
eat plants, but animals
also eat other animals
and compete for
resources. This creates
a complex food
web (see Food webs
and biomass, p.66)

**SEE ALSO**
Photosynthesis, p.58
Biodiversity, p.63
Agriculture, p.162

Nobody is as good at recycling as our entire planet. Important **nutrients** such as carbon, oxygen, nitrogen, phosphorus, and several others constantly cycle between living **organisms** and soil, rocks, air, or water.

For example, plants take up nutrients from soil that can be transferred to animals, who eat the plants, and later re-enter the soil when dead animals are decomposed.

As an example, let's look at the nitrogen cycle. Plants and animals need nitrogen to form **DNA**, **RNA** and **proteins**. Nitrogen is abundant in the atmosphere as $N_2$ gas, but most organisms can't use that directly. However, some bacteria turn atmospheric nitrogen into nitrogen **compounds** that plants can take up. This process is called nitrogen fixation. Animals then access the nitrogen by consuming plants. During decomposition of dead organisms, nitrogen re-enters the soil, from where it is eventually converted back into nitrogen gas and re-enters the atmosphere.

Chemical elements such as carbon, oxygen, nitrogen and phosphorus are essential to life. These nutrients continuously cycle between living organisms and non-living components such as air or soil. Several pathways and mechanisms are involved in each nutrient cycle. For example, **photosynthesis** and **respiration** are part of carbon and oxygen cycles. Bacteria and fungi play an important role in several nutrient cycles by converting nutrients into other forms. One of the cycles where this happens is the nitrogen cycle. Here, microbes carry out nitrogen fixation to turn atmospheric nitrogen into compounds that plants take up from soil and use to grow.

The nitrogen cycle constantly moves nitrogen between the atmosphere, soil and organic matter such as plants and animals.

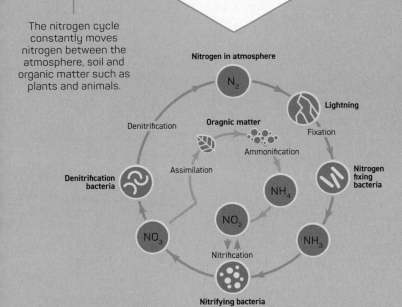

Nitrogen in atmosphere

$N_2$

Lightning

Denitrification

**Oragnic matter**

Ammonification

Fixation

Assimilation

Nitrogen fixing bacteria

**Denitrification bacteria**

$NH_4$

$NO_2$

$NH_3$

$NO_3$

Nitrification

**Nitrifying bacteria**

# 33

# Food webs and biomass

**WHY IT MATTERS**
Food webs are part of
nutrient cycles that
maintain balance
within an ecosystem

**KEY THINKERS**
Charles Elton
(1900–1991)
Raymond Lindeman
(1915–1942)

**WHAT COMES NEXT**
Food webs illustrate
how different species
live together in an
ecosystem. Some
species hunt each
other; some help
each other out (see
Living in groups, p.68
and Symbiosis, p.70)

**SEE ALSO**
Photosynthesis, p.58
Nutrient cycles, p.64
Nutrition and
digestion, p.126

Predators at the top of the food chain eat animals below them, which in turn eat other animals or plants.

In reality, there are many interlinking food chains, because animals eat different types of food and several **species** compete for the same food source. Instead of a chain, it's a food web.

Each level of the food web is a trophic level. The producers at the lowest trophic level make up the majority of the biomass. In ecology, biomass is a measure of the total amount of organic material or of how much energy is stored in that **organism**. Not all of the biomass is transferred to the next trophic level, so there is more biomass in plants than there is in apex predators.

"Without sharks, you take away the apex predator of the ocean, and you destroy the entire food chain."

**Peter Benchley,** author of *Jaws*

Animals eat other animals (or plants) from a lower level of the food web. At the lowest trophic level are producers or autotrophs. These are plants or algae that make their own food through **photosynthesis**. They're eaten by primary consumers, which in turn are eaten by secondary consumers. Marine food webs usually have more levels than land food webs.

Not all biomass from lower trophic levels is transferred to consumers at higher levels, because most of it is lost as heat or waste. So in any **ecosystem**, most biomass is in producers, and biomass decreases with each higher trophic level.

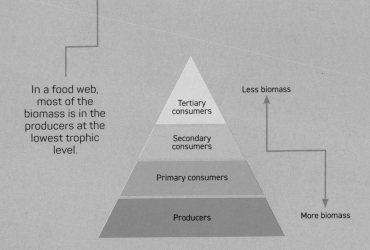

In a food web, most of the biomass is in the producers at the lowest trophic level.

Less biomass

Tertiary consumers

Secondary consumers

Primary consumers

Producers

More biomass

# 34

# Living in groups

Birds of a feather flock together, but so do other animals. Living in groups is common across many **species** because it provides safety in numbers, shared resources, better food gathering strategies, social contact and more.

Animals that live in groups have found ways to communicate with each other, whether that's through sounds, gestures, scent or another method. Downsides to group living include competition for resources and mates, being more visible to predators and being more susceptible to infectious diseases. Some animals, such as bears, sea turtles and leopards, do not live in groups and prefer a more solitary lifestyle.

**WHY IT MATTERS**
The way animals live in groups shows how they interact with the rest of their ecosystem and it informs conservation efforts

**KEY THINKERS**
Bees: Karl von Frisch (1886–1982)
Gorillas: Dian Fossey (1932–1985)
Chimpanzees: Jane Goodall (1934–)

**WHAT COMES NEXT**
Animals often live in groups together with other species (see Symbiosis, p.70)

**SEE ALSO**
Animal behaviour, opposite
Populations and ecosystems, p.62
Epidemics and pandemics, p.76

In
**100**
words

Many animals live in large groups with other members of the same species. There are different reasons why animals might stick together. For example, to offer protection from predators, to help each other find food or to find a mate. Animals that live in groups often develop ways to communicate important information to each other. For example, bees do a dance in the air to let other bees know they found food, while primates use gestures and facial expressions to communicate.
One of the downsides of living in groups is that it makes it easier for infectious diseases to spread.

## In 100 words

Ethologists are biologists who study animal behaviour to learn how different **species** act, communicate and learn in their usual surroundings. Understanding this behaviour shows what's important for these species to survive. For example, Konrad Lorenz discovered that if baby geese don't see their mother soon after hatching, they will treat another species or even an inanimate object as their mother instead. Observing animal behaviour in their natural environment highlights how different some animals act when they're in captivity. This type of research has led to better treatment for animals in captivity, such as livestock for agriculture or animals in zoos.

# Animal behaviour

Why do animals do the things they do? To answer this question, researchers have been observing animal behaviour in the wild. They've found that animals can have many different ways of finding a mate, raising their young, communicating, hunting or resting. For example, some animals use tools to find food.

The study of animal behaviour, or ethology, is important for animal welfare, for conservation and for understanding human behaviour.

**WHY IT MATTERS**
Animal welfare and conservation policies rely on understanding what regular behaviour looks like to be able to spot when something is wrong

**KEY THINKERS**
Konrad Lorenz (1903–1989)
Nikolaas Tinbergen (1907–1988)
Temple Grandin (1947–)

**WHAT COMES NEXT**
Some forms of animal behaviour are meant as methods of communication with other members of their population (see Living in groups, opposite)

**SEE ALSO**
Populations and ecosystems, p.62
Sexual selection, p.84
Agriculture, p.162

# 36

# Symbiosis

Clownfish and anemones are never far apart. The anemone provides shelter for the clownfish and the clownfish provides **nutrients** for the anemone. This intertwining of two different **species'** lives is called symbiosis, and it is common between many species. We're all in this together! Symbiosis isn't always to the benefit of both species. For example, a parasite that takes advantage of another **organism** is also practising a form of symbiosis.

**WHY IT MATTERS**
Symbiosis means that species influence each other within their ecosystem

**KEY THINKERS**
Symbiosis between humans and the environment: Indigenous people in the Americas (since c. 1000 BCE)
The term symbiosis: Albert Bernhard Frank (1839–1900) and Heinrich Anton de Bary (1831–1888)
Endosymbiosis: Lynn Margulis (1938–2011)

**WHAT COMES NEXT**
Symbiosis can explain how different organisms evolve (see Evolution, p.18). The effects of symbiosis also affect agriculture (see p.162), human health (see Microbiome, p.74), and our understanding of the spread of diseases

**SEE ALSO**
Evolution, p.18
Populations and ecosystems, p.62
Conservation, p.163

## In 100 words

Symbiosis describes a situation where two different species live closely together. Often, symbiosis refers to mutualism, where both species benefit. But there are other forms of symbiosis where only one species benefits while the other is neutral about the collaboration (commensalism), or where one organism takes advantage of another (parasitism). If one species lives inside another, this is called endosymbiosis. Because species influence and depend on each other in many ways, symbiosis influenced the way biologists think about conservation, agriculture, health and disease. Endosymbiosis of bacteria living inside primitive eukaryote cells can even explain the evolution of mitochondria and chloroplasts.

## In 100 words

Mycorrhizae are symbiotic relationships between plants and fungi that grow on and inside of the roots of the plant or tree. The fungi have thread-like roots called mycelia (singular: mycelium) that extend deep into the soil. Mycelium filaments can form connections between trees, but how large or influential those networks are is still being researched.

The large mycelium branches that spread out from fungi help decompose organic material in the soil so that nutrients can be made available again for other organisms. Mycorrhizal fungi take up large amounts of space underground, and they are important for keeping ecosystems in balance.

# Mycorrhizae

**Ecosystems** extend to the air and soil that **organisms** need to grow. Plants and trees take up **nutrients** from soil, but they've got some help. Fungi that grow on tree roots form symbiotic connections called mycorrhizae in which the fungus helps the tree absorb nutrients from soil and gets sugars in return.

Mycorrhizae can also form networks between trees, but biologists are still debating how extensive or important these networks are. What's certain, though, is that fungi are crucial to forest ecosystems.

## WHY IT MATTERS
Even though they're hidden underground, mycorrhizae play an important role in the ecosystem by providing nutrients for plants

## KEY THINKERS
Albert Bernhard Frank (1839–1900)
Suzanne Simard (1960–)
Melanie Jones (c. 1960–)
Toby Kiers (1976–)

## WHAT COMES NEXT
The mycelia of mycorrhizal networks help decompose dead organic material to release nutrients back into the soil (see Nutrient cycles, p.64)

## SEE ALSO
Plants, p.14
Photosynthesis, p.58
Biodiversity, p.63
Food webs and biomass, p.66

# 38

# Coral reefs

## WHY IT MATTERS
Coral reefs are some of the most biodiverse areas in the sea. They also protect the coast from erosion, and they're economically important to nearby countries through fishing and tourism industries

## KEY THINKERS
Peter Glynn (1933–)
Terry Hughes (1956–)
Nyawira Muthiga (1957–)

## WHAT COMES NEXT
To mitigate the threats of climate change and other coral bleaching events, coral reefs are a key focus of conservation (see Conservation, p.163)

## SEE ALSO
Invertebrates, p.16
Biodiversity, p.63
Food webs and biomass, p.66

Coral reefs are marine **ecosystems** with a lot of biodiversity. They're home to a complex food web from plankton to reef sharks with many **species** taking advantage of the shelter and nearby food sources that corals provide. Even though coral reefs take up less than 1 per cent of the ocean floor, they're home to about a quarter of marine life.

Coral reefs are usually found close to the coast in or near tropical regions. They protect the coastal area from erosion and are crucial to the fishing and tourism industries of the countries near the reefs. Combined with the high biodiversity levels of reefs, this makes them worth protecting. However, imbalances such as high water temperatures can throw the entire ecosystem off. It can cause coral polyps to lose the algae that provide them with **nutrients** and energy, which makes the coral more susceptible to disease and more likely to die.

**In 100 words**

Corals are invertebrates in the phylum Cnidaria. Identical coral polyps grow close together to form corals of different shapes and colours. The colour comes from single-celled algae called zooxanthellae that live within the polyps and provide them with nutrients. Corals are primary consumers in the marine food web: they eat plankton and are eaten by fish. Corals also offer shelter and protection, which attracts animals to the coral reef and creates a biodiverse ecosystem.

High water temperature or other factors can cause corals to lose their algae. This coral bleaching puts the coral at increased risk of death or disease.

Many individual coral polyps cluster closely together in colonies.

## CORAL ANATOMY

Tentacles

Mouth

Nematocyst

Mesogloea

Zooxanthellae

Pharynx

Septum

CaCo$_3$ Skeleton

# 39

# Microbiome

## WHY IT MATTERS
Not all bacteria are pathogens. In the human body, they help digest food and train the immune system

## KEY THINKERS
Faecal transplantation to treat gut disorders: Li Shizhen (1518–1593), Ge Hong (283–343)
First to see microbes on human body: Antonie van Leeuwenhoek (1632–1723)
Gut microbiome: Alfred Nissle (1874–1965) and Jeffrey Gordon (1947–)
Bobtail squid and microbes: Margaret McFall-Ngai (c. 1951–)

## WHAT COMES NEXT
Researchers use genetics to identify microbes that live in the human body (see Genetic sequencing, p.156)

You're never truly alone. The human body – and that of many other animals – is host to trillions of bacteria at all times. This **ecosystem** of bacteria within us is called the microbiome, and the gut microbiome is specifically the collection of all bacteria that live in the digestive system.

Gut bacteria help to digest food, but they also prevent disease-causing bacteria from taking over. Sometimes this balance can be thrown off. Even as early as the fourth century, physicians in China knew that they could treat gut disease by transferring faeces from a healthy person. They didn't yet know that this had to do with the gut microbiome, but from the earliest microscope studies in the 17th century, biologists have known that bacteria live in and on us.

More recently, researchers have been learning about all the different roles of the microbiome, from digesting food to helping to develop the immune system. Other animals have their own microbiomes, such as the Hawaiian bobtail squid, which gets a special light organ thanks to bacteria.

In
**100**
words

Most animals, including humans, have
trillions of bacteria living in and on their bodies.
Many of these bacteria are largely harmless or have
useful functions within the body, such as digesting food
or training the immune system to distinguish harmless bacteria
from dangerous ones. Healthy gut bacteria also help fight disease-
causing bacteria. Some digestive diseases caused by bacterial
**pathogens** can be treated by introducing healthy gut
bacteria, such as by transplanting faeces
from a healthy person. Other animals also have
symbiotic relationships with bacteria.
The Hawaiian bobtail squid has
bioluminescent bacteria that
produce light which hides
the squid from
predators.

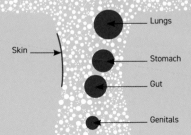

Nose
Mouth

Lungs

Several areas of
the human body
are host to their
own microbiomes.

Skin

Stomach

Gut

Genitals

# 40

# Epidemics and pandemics

## WHY IT MATTERS
Human health and food supply can both be affected by infectious diseases that spread through the population

## KEY THINKERS
First documented plague: Galen (129–216)
Epidemiology: John Snow (1813–1858)
Genetic diversity in plants protects against disease: Norman Borlaug (1914–2009)

## WHAT COMES NEXT
At an individual level, the human immune system has a defence mechanism for any infection that enters the body (see Immune system, p.116)

## SEE ALSO
Living in groups, p.68
Agriculture, p.162
Germ theory, p.164
Vaccines, p.166

Throughout history, humans have been affected by epidemics and pandemics. Some diseases, such as the bubonic plague, Spanish flu, AIDS and COVID-19 are spread between people or jump from animals to people. Others, such as cholera, were linked to contaminated food or water sources. Epidemics can also affect humans indirectly when they occur in livestock or agricultural crops. For example, the potato famine in 19th-century Ireland was caused by the potato blight fungus.

Not all **organisms** will react in the same way to an infection. Some may get sick and die, some get sick and pass on the **pathogen**, some pass it on without getting sick, and others aren't affected at all. There is variation between **species**, but also among genetically different members of the same species. That's why a lack of **genetic diversity** in agriculture can put an entire crop at risk.

Epidemiology is the study of how diseases are spread and how they can be contained. Some solutions that have been known since at least the Middle Ages were to isolate the source of infection and prevent further spread. Since the 18th century, vaccination has been another way to reduce the impact of epidemics.

The rapid outbreak of an infectious disease in a **population** is an epidemic. If it spreads to a global level, it becomes a pandemic. Outbreaks can be contained by removing the source of an infection, blocking the way the pathogen spreads, or through vaccination to improve the immune response of individuals and populations. A genetically diverse population will be more resilient, because not all individuals will be equally susceptible to the disease. This is true for any organism, but is particularly important for agriculture, where genetically identical crops may be at more risk of being wiped out by an infection.

"Epidemics on the other side of the world are a threat to us all. No epidemic is just local."

**Peter Piot,** Ebola and AIDS researcher

# Life Cycles

The birds and the bees – as well as every other organism – have to reproduce to keep their species going. Some use asexual reproduction, others need both female and male parents to combine their genetic material, but all of them have found a way to create new life.

Multicellular organisms start their lives by developing specialised cells and organs, then growing and maturing until they're capable of producing their own offspring.

Eventually, death is inevitable, but biologists are studying what happens to cells and DNA during ageing so that people can remain as healthy as possible even later in life.

# 41

# Mitosis and meiosis

**Reproduction** is simple for **single-celled** organisms: they just divide to create more of them. For larger **organisms**, things get more complicated. Many plants and animals use sexual reproduction, for which they need to combine their **DNA** without duplicating the total number of chromosomes. And after that, any multicellular organism needs to grow from a single **cell** to many specialised cells.

But whether you grow a tree or an elephant, the cellular processes are the same. All eukaryotes use mitosis to form new cells during growth. Organisms that use sexual reproduction also use a process called meiosis to form **gametes** (egg and sperm cells).

Mitosis and meiosis both start by duplicating DNA so that each chromosome has two identical halves.

During mitosis, these halves are pulled apart to opposite sides of the cell. The cell then divides into two daughter cells that each have the same **genetic** material.

During meiosis, the chromosomes first form a pair with their homologous chromosome. These pairs are then separated when cells divide, and further split into chromatids during the next **cell division**.

# In 100 words

During mitosis, a cell splits into two daughter cells that each have the same DNA as the parent cell. Multicellular organisms use mitosis as they grow during **development** and to replace cells. Single-celled organisms use mitosis to reproduce. Meiosis is a different kind of cell division that results in gametes (or sex cells), which each have only one copy of each chromosome (haploid cells) instead of two (diploid cells). The chromosomes can be divided between daughter cells in different ways so that not all resulting cells are identical. When these cells combine during reproduction, they combine chromosomes from both parents.

Mitosis: In mitosis, the DNA of a diploid (2n) cell first doubles (4n) before the cell splits off into two daughter cells that each have the original diploid genome.

Meiosis: In meiosis, a diploid (2n) germ cell first doubles its DNA (4n) before pairing off into daughter cells (2n), which are then further split into the haploid (n) cells known as gametes.

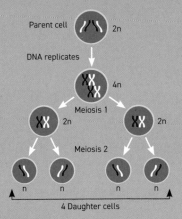

Parent cell — 2n

DNA replicates

4n

2n — 2 Daughter cells — 2n

Parent cell — 2n

DNA replicates

4n

Meiosis 1

2n — 2n

Meiosis 2

n — n — n — n

4 Daughter cells

# 42

# Plant development

**WHY IT MATTERS**
Understanding how plants grow and develop is crucial to agriculture

**KEY THINKERS**
Seed to plant: Mesopotamians (c. 10,000–2000 BCE)
Plant fertilisation: Joseph Gottlieb Kölreuter (1733–1806)
Arabidopsis: Friedrich Laibach (1885–1967)
Plant hormones: Ottoline Leyser (1965–)

**WHAT COMES NEXT**
Gregor Mendel cross-bred plants to study how different traits were inherited from one generation to the next (see Inheritance, p.86)

Understanding how to grow plants from seeds was something the first farmers already understood thousands of years ago. Since then, biologists have learned a lot more about plant development.

Much of the genetics and biology of plant development is being studied in the model **organism** *Arabidopsis thaliana*. This is a flowering plant that's easy to grow and research.

Plants grow differently from animals because they're fixed to the same location throughout their lifecycle. However, they can adapt quite well to different conditions in their environment — for example, by growing bigger roots to find more **nutrients**.

Another difference with animals is that plant cells have **cell** walls and a water-filled **organelle** called a vacuole. This allows them to use a process called cell elongation, in which the cells change shape to help the plant grow.

All flowering plants grow from seed. Their flowers produce new seeds to grow the next generation.

### In
# 100
### words

Plants use either seeds or spores to reproduce. Flowering plants produce seeds through sexual **reproduction**, where pollen from one flower lands on the stigma of another flower to start the process of seed formation.

Seeds already have everything they need to develop into a mature plant, as long as they receive water, soil and sunlight.

As plant seeds germinate, they grow roots, stems and leaves. Plant growth is regulated by growth regulators, such as the hormone auxin, but even gravity plays a role in helping plants grow upright. In addition to **cell division**, plants also use cell elongation to grow.

# 43

# Sexual selection

Why do male peacocks have beautiful tails and male deer such huge antlers? The answer is sex. Both animals make a big show of themselves to attract a mate. The deer also use their antlers to compete with each other over a female deer.

This type of competition is called sexual selection, and it's a form of natural selection that helps a **population** to carry on **genetic traits** to the next generation.

In a way, even flowering plants show a form of sexual selection, because pollinators may prefer certain flowers.

**WHY IT MATTERS**
Sexual selection is a form of natural selection by competition for a partner

**KEY THINKERS**
Charles Darwin (1809–1882)
Angus Bateman (1919–1996)
Helena Cronin (1942–)

**WHAT COMES NEXT**
After animals have found their sexual partner, they can reproduce to pass on their genetic traits (see Animal reproduction, opposite)

**SEE ALSO**
Evolution, p.18
Populations and ecosystems, p.62
Living in groups, p.68

## In 100 words

Sexual selection describes how animals find a partner to mate with. Darwin first described two forms of sexual selection: intrasexual selection, where male individuals compete with each other for access to a female, and intersexual selection, where females choose between male partners. This competition exists because the female individuals in most populations only have a few eggs available while the males have many sperm cells with which they can produce offspring. Some animal characteristics only make sense in the context of sexual selection. For example, male peacocks' elaborate tails attract female birds, but also make them more visible to predators.

## In 100 words

Most animals reproduce by fertilising a female gamete (egg) with a male gamete (sperm). However, some invertebrates, amphibians and reptiles can reproduce asexually with only one parent. For mammals and most egg-laying **species**, **fertilisation** occurs inside the body. Fish lay eggs before they are fertilised. Only mammals develop entirely in the womb, while other animals develop in eggs outside the body. Since egg and sperm cells only have one set of chromosomes (haploid), combining them through fertilisation results in a diploid **cell** with chromosome pairs. This **zygote** cell further develops into an embryo, foetus and eventually into a mature organism.

# Animal reproduction

Animals that use sexual reproduction need to combine two **gametes** – egg and sperm – to start forming a new **organism**. The basic principles are the same for all these animals, but there are a few differences. For example, fish lay their eggs before the male fish fertilises them, while for other egg-laying animals and mammals, fertilisation takes place inside the body.

In almost all mammals, the fertilised egg further develops inside the womb. The exception is monotremes (the platypus and echidnas), which are mammals that lay eggs.

**WHY IT MATTERS**
Fertilisation of an egg by sperm during sexual reproduction marks the start of embryonic development

**KEY THINKERS**
Charles Bonnet (1720–1793)
Oskar Hertwig (1849–1922)
Ernest Everett Just (1883–1941)

**WHAT COMES NEXT**
It is also possible to fertilise mammalian eggs outside of the body in a lab – for example, to help people who struggle with infertility (see IVF, p.170)

**SEE ALSO**

# Inheritance

Long before anyone knew about **DNA** and **genes**, Gregor Mendel used pea plants to figure out that the information for inherited **traits** comes in pairs and that these traits, or **phenotypes**, are either dominant or recessive.

An example of this is eye colour. Having blue eyes is a recessive trait, while brown eyes are dominant. Someone with blue eyes (b) will always have two copies of the blue eyes **allele** (bb). They're homozygous for that trait. But people with brown eyes (B) can be either homozygous (BB) or heterozygous (Bb) because the brown eye allele is dominant.

Punnett squares show how inheriting different **genotypes** can lead to observable phenotypes. For example, it shows why there is a 25 per cent chance for any child of two heterozygous carriers of a recessive trait to have the trait, or why children of a homozygous and heterozygous parent have a 50 per cent chance to also be heterozygous.

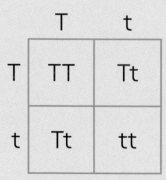

two heterozygous parents Tt x Tt

This Punnett square shows that a child of parents who are both carriers (Tt) of a recessive trait t has a 25% chance of inheriting the recessive allele t from both parents (tt), a 50% chance of becoming a carrier themselves (Tt) and a 25% chance of not carrying the recessive gene at all (TT).

**Genetic** material from both parents combines during **fertilisation**. Diploid **species** (such as humans) may have different versions of genes on each chromosome of a pair. These versions are called alleles. Recessive traits express if both chromosomes have the recessive allele. They have a homozygous genotype with the same allele on both chromosomes. Dominant traits can have either homozygous or heterozygous genotypes. Most human male (XY) individuals only have one X chromosome, so X-linked recessive traits affect men more than women. Dominant, recessive and X-linked inheritance are referred to as Mendelian inheritance, whereas mitochondrial DNA is always inherited from the mother.

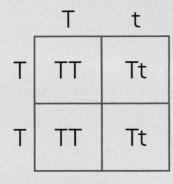

|   | T | t |
|---|---|---|
| T | TT | Tt |
| T | TT | Tt |

homozygous and heterozygous
TT x Tt

This Punnett square shows that a child of a parent with two dominant alleles (TT) and a heterozygous parent (Tt) will always show the dominant trait T. They have a 50% chance of being heterozygous (Tt) and a 50% chance of being homozygous dominant (TT).

# 46

# Embryonic development

**WHY IT MATTERS**
Embryonic
development forms
the early stages
of an animal's life,
after which it
further develops
as a larva or foetus.

**KEY THINKERS**
Marcello Malpighi
(1628–1694)
Karl Ernst von Baer
(1792–1876)
Hans Spemann
(1869–1941)
Hilde Mangold
(1898–1924)

**WHAT COMES NEXT**
Development is
regulated by genes
that are conserved
across species
(see Genes in
development, p.88)

**SEE ALSO**
Vertebrates, p.17
Model organisms, p.20
Animal
reproduction, p.85
Stem cells and cell
differentiation, p.96

Very early in **development**, all animals look pretty similar. After **fertilisation**, they all follow the same route, where the **zygote** divides and the resulting mass of **cells** folds into shapes that will eventually become the embryo.

After the embryo is formed, it moves on to the next developmental stage. Some animals, such as insects or amphibians, hatch into a larva which undergoes metamorphosis to become a mature **organism**. Others, such as birds and mammals, move from the embryonic stage to the foetal stage. In humans, embryonic development from zygote to foetus takes about 8 weeks.

Because embryonic development is so similar for many **species**, researchers can learn about human development by studying model organisms.

In the first steps of embryonic development a fertilised egg (zygote) divides into multiple cells. These cells reorganise and form germ layers that will eventually become tissues and organs.

Fertiised egg    2 Cell stage    4 Cell stage    8 Cell stage

In **100** words

Embryonic development describes the steps that happen after a sperm and egg cell have combined to form a zygote.

The zygote first divides several times until it is a clump of smaller cells. These cells rearrange to form a hollow sphere called a blastula, which has a fluid-filled centre. In mammals, the blastula stage is called a blastocyst. It has an inner cell mass that will become the embryo.

The blastula (or blastocyst) changes shape and becomes a gastrula, which has folds that form three germ layers (endoderm, ectoderm and mesoderm) which are the origins of different **tissues** and organs.

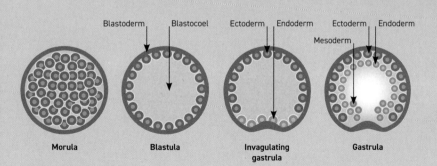

Blastoderm    Blastocoel    Ectoderm    Endoderm    Ectoderm    Endoderm

Mesoderm

Morula          Blastula          Invagulating          Gastrula
                                   gastrula

# 47

# Genes in development

## WHY IT MATTERS
Genes expressed during early development determine the basic body plan of organisms, such as where the head and legs go, as well as which sex organs are formed during development

## KEY THINKERS
Hox genes: the research groups of Walter Gehring (1939–2014) and Thomas Kaufman (c. 1944–)
SRY gene: the research groups of Robin Lovell-Badge (1953–) and Peter Goodfellow (1951–)

## WHAT COMES NEXT
Some genes influence other genes by activating or inhibiting their expression (see Gene expression, p.35). But gene expression can also be regulated by factors outside of the genome (see Epigenetics, p.92)

## SEE ALSO
Chromosomes, p.32
Genes, p.34
Embryonic development, p.88

Even though each embryo starts out as a small ball of **cells**, it manages to develop complicated body parts and organs. This process is partly orchestrated by **genes** that express at the right time and the right place, such as Homeobox (Hox) genes, SRY and several others.

Hox genes were first discovered in flies, but all animals have them. They're close together in the **genome** and they're expressed in a very coordinated manner during embryonic **development** so that each body part, such as the head or limbs, ends up in the right place.

The SRY gene is found on the Y chromosome. If this gene is expressed, the foetus usually develops male sex characteristics. However, if the SRY gene is mutated or changes location, it may lead to disorders of sex development (for example, XY with female sex characteristics).

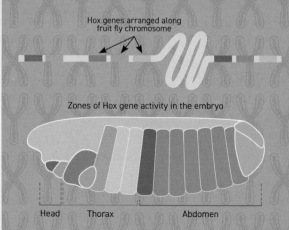

Hox genes arranged along fruit fly chromosome

Zones of Hox gene activity in the embryo

Head        Thorax                    Abdomen

In
# 100
**words**

The Hox genes and the SRY gene are some of the important **genetic** factors that coordinate embryonic and foetal development. Hox genes are expressed in a coordinated manner as the embryo develops, and their activity leads to the formation of key parts of the body such as the head and limbs. The SRY gene is present on the Y chromosome and is expressed during development. It determines which sex organs form in the foetus. Usually, male sex characteristics only develop if SRY is expressed. There are other genes involved in development as well, such as the Fox and Sox genes.

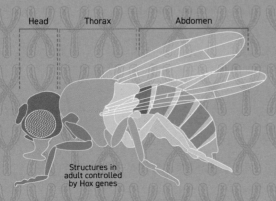

Head        Thorax                    Abdomen

Structures in
adult controlled
by Hox genes

Hox genes in fruit flies determine where each body part is formed. The order of Hox genes in the fly's DNA is directly related to the order in which the corresponding body parts appear in the fly.

# 48

# Epigenetics

How **genes** act is not entirely determined by genetics alone. Diet, stress, exercise and other factors can all lead to **epigenetic** changes.

Two examples of epigenetic control are **DNA** methylation and histone modification, which both change the way that certain genes are expressed.

Epigenetics explains, for example, why identical twins (who have the same DNA) can still be slightly different.

Epigenetics affects **development** in a few ways. For example, even though humans inherit two **alleles** of most genes, they're not always expressed in equal amounts. Some genes are expressed mainly from the chromosome inherited from the father or that from the mother. This is called genomic imprinting, and it's controlled by epigenetics.

Throughout life, environmental factors can continue to change the way that genes are expressed. Some epigenetic factors can even pass from one generation to the next if they affect mothers during pregnancy. However, unlike changes in the DNA itself, epigenetic changes are reversible.

**WHY IT MATTERS**
Epigenetics explains how changes in environment influence the way that genetic information is expressed

**KEY THINKERS**
Epigenetics:
Conrad Waddington
(1905–1975)
Genomic imprinting:
Azim Surani (1945–)
DNA methylation:
Robin Holliday
(1932–2014),
Arthur Riggs
(1939–2022)
Histone modification:
Vincent Allfrey
(1921–2002),
David Allis (1951–2023)

**WHAT COMES NEXT**
Epigenetics also controls how stem cells become different cell types (see Stem cells and cell differentiation, p.96)

**SEE ALSO**
Gene expression, p.35
Mutation and variation, p.42

Epigenetics describes how genes are controlled by factors other than genetics. Two examples are DNA methylation and histone modification.

In DNA methylation, a small molecule (methyl) is attached to the DNA, which affects how the gene at that location is expressed. Histone modification changes the histone **proteins** that DNA is wound around to form chromatin and chromosomes. These modifications can make it easier or more difficult for a gene to be expressed.

Whether these epigenetic changes take place can depend on environmental factors such as diet, exercise or exposure to stress. Epigenetics plays a role in development and throughout life.

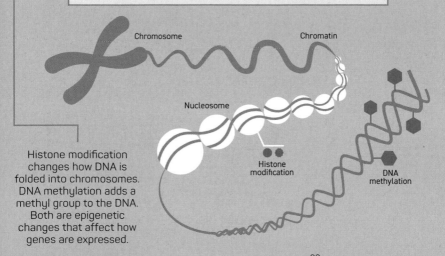

Chromosome

Chromatin

Nucleosome

Histone
modification

DNA
methylation

Histone modification changes how DNA is folded into chromosomes. DNA methylation adds a methyl group to the DNA. Both are epigenetic changes that affect how genes are expressed.

93

# 49 Growth and growing up

Most animals are not yet at full size when they're born or hatched, but not all **species** take the same amount of time to go from baby to adult. Like many other animals, humans go through childhood and puberty before they become adults. However, neuroscientists have discovered that human brains continue developing for a few more years after that, so it's difficult to pinpoint exactly when someone has finished growing and matured into an adult.

**WHY IT MATTERS**
Development during childhood and adolescence sets the stage for later life

**KEY THINKERS**
Growth chart:
Henry Bowditch
(1840–1911)
Growth plate:
John Hunter
(1728–1793)
Gonadotropin-releasing hormone:
Roger Guillemin
(1924–) and
Andrew Schally (1926–)
Adolescent brain:
Sarah-Jayne Blakemore
(1974–)

**WHAT COMES NEXT**
Despite an early growth spurt in childhood, our bodies don't keep growing forever. There is a limit to how tall you can get, and that limit varies a lot for different organisms (see Variation in size between species, opposite)

**SEE ALSO**
Animal behaviour, p.69
Hormones, p.114
Brain and nervous system, p.124

## In 100 words

Animals get bigger when they acquire more **cells** through **cell division**. Growth plates at the end of vertebrates' bones makes these grow longer until adult height is reached. Paediatricians and parents often track whether children fall within the average growth range during childhood by measuring height and weight. Puberty marks the transition from childhood to adulthood. Hormone levels change, starting with a rise in levels of gonadotropin-releasing hormone. Puberty further develops based on the relative presence of different sex **hormones**. Even after becoming fully grown sexually mature adults in their late teens, humans' brains continue to develop into their twenties.

In **100** words

There are limits to how large an animal can get. Insects are usually small, because their **exoskeleton** would make it too difficult to breathe if they were larger. Animals with a more efficient oxygen exchange system can get bigger. Invertebrates and mammals in the ocean tend to be larger than those on land, because the buoyancy of the water helps to support their weight.
Climate, diet and food availability also play a role, and help explain why some animals were larger in prehistoric times, or why animals on islands can be unusually large or small in more strictly controlled **ecosystems**.

# 50

## WHY IT MATTERS
Understanding the limits on animal sizes can help scientists predict what will happen when climate or prey availability changes. Some groups of animals may get bigger or smaller over time

## KEY THINKERS
J. B. S. Haldane (1892–1964)
J. Bristol Foster (c. 1933–)

## WHAT COMES NEXT
Big or small, some animals live for many years, but they don't keep the same cells the entire time (see Cell death and renewal, p.100)

## SEE ALSO
Evolution, p.18
Biodiversity, p.63
Food webs and biomass, p.66
Homeostasis, p.130

# Variation in size between species

Animals come in all shapes and sizes. Ants are always small, while elephants or whales are large.

Physics can explain why large insects would spend too much energy breathing and why whales could never support their own weight on land. Other limitations have to do with climate, food availability, changes in diet or the presence of predators. All these factors and others have put evolutionary pressure on animals to become a certain size. Over long periods of time, this can change again.

# 51

# Stem cells and cell differentiation

## WHY IT MATTERS
All cells, tissues and organs start out as stem cells before they become specialised through differentiation

## KEY THINKERS
Embryonic stem cells: Martin Evans (1941–)
Haematopoietic stem cells: Ernest McCulloch (1926–2011) and James Till (1931–)

## WHAT COMES NEXT
Some medical conditions can be treated with stem cell therapies, such as a bone marrow transplant that provides new haematopoietic stem cells (see Stem cell therapy, p.174)

## SEE ALSO
Gene expression, p.35
Cells, p.26
Cell signalling and transport, p.50
Epigenetics, p.92
Tissues and organs, p.106

Even though embryos start out as just a few cells, mature multicellular **organisms** have many different cell types. That's because embryonic stem **cells** are able to divide into new cells that take on different specialisations depending on the signals they receive from surrounding cells and their environment. These signals regulate which **genes** are expressed in the new cell, and that determines the cell's function.

Stem cells in the early embryonic stage can turn into any cell type. Adults also have stem cells, but those can only turn into a particular group of cells. For example, haematopoietic stem cells in bone marrow can become all the different types of blood cells.

Stem cells have the potential to become a more specialised cell type. Embryonic stem cells that can still become any type of cell in the developing organism are called **pluripotent**. Adult stem cells can either turn into several different cell types or just a single cell type as part of the natural turnover of cells in the body. Stem cells divide into more stem cells and some of these new cells will change into specific cell types. That process is called differentiation and it's regulated by cell signalling pathways and **epigenetic** factors that determine what these new cells turn into.

Stem cells have the potential to differentiate into several types of specialised cells.

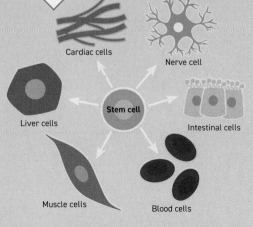

Cardiac cells

Nerve cell

Liver cells

Stem cell

Intestinal cells

Muscle cells

Blood cells

# 52

# Cancer

Every year, 18 million people around the world are diagnosed with one of the many different types of cancer. Cancers can affect different parts of the body and have different causes and different treatments. What they have common is that they're characterised by cells growing out of control as a result of mutations in their **DNA**.

Some cancers are hereditary, where a parent passes on a **genetic** mutation that increases the risk of developing a certain cancer. But cancer-causing mutations in DNA can also be acquired in other ways — for example, through smoking or UV radiation from too much sun. Not every mutation forms a tumour, and not every tumour becomes cancer, but with every new mutation, the cancer risk increases.

Even though all cancers are different, the traditional treatment options are broadly similar for all cancers. Radiotherapy treats cancers using radiation. Chemotherapy attacks **cells** during the process of **cell division** (which happens more often in cancer cells). However, newer methods such as immunotherapy or precision medicine are now becoming available to treat some cancers in a more targeted way.

Most cancers are easier to treat if they are detected earlier. That's why many countries regularly screen healthy adults for some common cancers such as cervical cancer.

Mutations in DNA can cause cells to become cancer cells. Many cancer-causing mutations are found in proto-**oncogenes**. In healthy cells, these genes play a role in many regular processes such as cell division, cell differentiation or **apoptosis**. When a mutation disrupts these pathways, cells might divide too often and form a mass of cells, or tumour. If cancer cells spread from a tumour to other locations in the body, this is called metastasis. General treatment options for cancer include radiation therapy and chemotherapy, but for some cancers, newer methods are being developed that have fewer side effects and better outcomes.

"Every cancer looks different. Every cancer has similarities to other cancers. And we're trying to milk those differences and similarities to do a better job of predicting how things are going to work out and making new drugs."

**Harold Varmus,** former director of the National Cancer Institute

## 53

# Cell death and renewal

**WHY IT MATTERS**
Controlled cell death through apoptosis makes it possible for some parts of the body to renew old cells and replace them with fresh new cells

**KEY THINKERS**
Carl Vogt (1817–1895)
Sydney Brenner (1927–2019)
John Sulston (1942–2018)
Robert Horvitz (1947–)

**WHAT COMES NEXT**
If apoptosis doesn't work properly, cells continue to multiply without old cells being removed. This is one of several known causes of cancer (see Cancer, p.98)

**SEE ALSO**
Cells, p.26
Cell signalling and transport, p.50
Growth and growing up, p.94

It might sound counterintuitive, but cell death is necessary to live. For example, early in **development**, human hands have skin between their fingers (like ducks have on their feet), which largely disappears before birth as a result of programmed **cell** death, or **apoptosis**.

Many cells in the body, such as those in the lining of the intestines, are replaced regularly. Apoptosis removes any cells that aren't needed anymore in a controlled manner and makes room for new cells.

> **"**Without programmed cell death, the bonds that bind cells in complex multicellular organisms might never have evolved.**"**
>
> **Nick Lane** in *Power, Sex, Suicide: Mitochondria and the Meaning of Life*

Apoptosis is the process by which cells are broken down after they are no longer needed in the body.

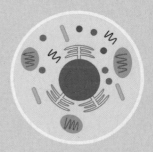

Normal cell

More than 50 billion cells in the body are renewed daily, requiring old cells to be removed as new ones take their place. This is done through a process called apoptosis, or programmed cell death. It's also relevant during development to remove cells that are no longer needed. Apoptosis can start from signals inside of the cell (intrinsic) or outside of the cell (extrinsic). Both routes activate several types of caspase **enzymes**. These destroy other **proteins**, which leads to the cell gradually breaking down. The remaining pieces of the dead cells are removed from the body by the immune system.

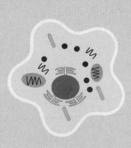

Shrinkage

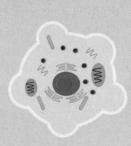

Membrane blebbing

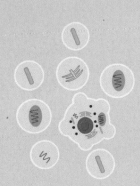

Cell breaks into apoptotic bodies

# 54

# Ageing

Even healthy individuals get older and eventually die. The average lifespan varies a lot between **species**. Flies and bees live only a few weeks, while a Greenland shark can be several centuries old. Humans are somewhere in between, with an average lifespan of 70–80 years.

Researchers are still trying to figure out all the factors that determine why and how we age, but some pieces of the puzzle are already in place. Mitochondria, **telomeres** on chromosomes, and **epigenetics** all play a role.

Although some researchers are interested in finding out how we can live longer lives, many biologists agree that it's more important to find out how we can live healthier when we do age.

Studies in mice showed that they stayed healthier longer if they ate less, but the full extent of the link between diet, epigenetics and ageing in humans is not clear yet.

**In 100 words**

When individuals age, their skin becomes thinner,
their bones more fragile and their muscles
weaker. All these ageing processes are the result
of several changes that take place over time in
the cells of the body.
One of these processes is the shortening of telomeres –
**DNA** at the tips of chromosomes. Without telomeres,
the **cell** can't divide properly anymore. Another
cause of ageing is that mitochondria lose some
of their function and start accumulating free
radicals, which can lead to further damage
in the cell.
In mice, diet can affect epigenetic
changes that cause mice to
stay healthy for longer.

**"As long as I stay healthy, the
numbers shouldn't matter. I don't feel
my age, I don't work my age, I don't
think my age, and hopefully, I don't
look my age!"**
**Dolly Parton,** at age 77

# Human Body

Biologists study all living organisms, but as humans we're particularly interested in ourselves. Understanding how our bodies function when they're healthy can help to explain what goes wrong during disease and how to treat that.

At the top level, bodies are organised into organs and organ systems, but none of them work independently. The human body is a complex system of many cells, tissues and organs that all act together to keep us alive.

# 55

# Tissues and organs

People have been fascinated with how our bodies function for a long time. Early anatomists in ancient Egypt and ancient Greece learned about organs through dissecting dead bodies, either for mummification or for medical research.

Once biologists were able to look more closely at sections of the body under the microscope, they learned that organs consist of different tissues, which in turn are made of **cells** that all connect to each other.

Altogether, tissues and organs create the organ systems that keep us alive.

## In 100 words

The human body is organised into tissues and organs. Cells with similar functions are grouped together as tissues. Within tissues, cells connect through interactions between **proteins** on the cell membranes of neighbouring cells. This process is called cell adhesion. Tissues include **epithelial** tissue, connective tissue, nervous tissue and muscle tissue. These are found throughout the body and combine to form organs. For example, the intestines are made up of all four tissue types. Groups of organs further connect and interact to form more complex organ systems throughout the body, such as the cardiovascular system with blood vessels and the heart.

## In
# 100
### words

Muscles connect to bones through tendons to form the musculoskeletal system that allows us to move. The **proteins** myosin and actin form fibres within muscle cells. When these fibres slide past each other in response to signals from the nervous system, muscle cells change shape. This causes the muscle to pull a bone to one side, which moves that part of the body. Muscle cells use **ATP** to reset to their relaxed state.

Skeletal muscles are voluntary because you move them at will. Other muscles are involuntary, such as cardiac muscle in the heart and smooth muscle in other organs.

**WHY IT MATTERS**
The musculoskeletal system of bones and muscles makes it possible to move

**KEY THINKERS**
Anatomy: Galen (129–216), Ibn Sina (980–1037), Andreas Vesalius (1514–1564)
Actin and myosin: Wilhem Kühne (1837–1900), Brunó Straub (1914–1996), Albert Szent-Györgyi (1893–1986)

**WHAT COMES NEXT**
Bone marrow on the inside of bones produces different types of blood cells (see Blood, p.110)

**SEE ALSO**
ATP, p.52
Brain and nervous system, p.124
Dissection, p.135

# Bone and muscle

Bones and muscles create the frame for the body and form protective spaces for organs in the abdominal cavity and the skull. But perhaps most importantly, they make it possible to move around.

Adult humans have 206 bones that connect to each other via ligaments and to muscle via tendons.

To move, the brain sends signals through the nervous system to the muscles. These then contract, which causes the movement. Muscle contractions happen at the level of muscle **cells**, where the proteins **myosin** and actin form fibres that slide past each other to pull the cell together.

# Lungs and breathing

Take a deep breath in. You just filled your lungs with air that includes an important ingredient for all the **cells** in the body: oxygen. The first step in getting oxygen where it needs to be is to transfer it to the blood, and that happens deep inside the air sacs of the lungs. The lungs also pick up carbon dioxide from the blood, which you remove from your body with every exhalation.

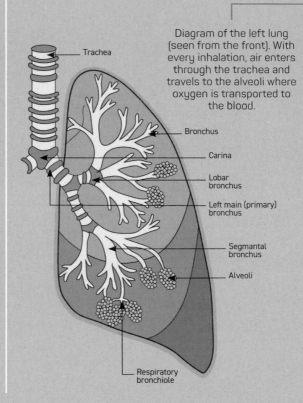

Diagram of the left lung (seen from the front). With every inhalation, air enters through the trachea and travels to the alveoli where oxygen is transported to the blood.

Trachea

Bronchus

Carina

Lobar bronchus

Left main (primary) bronchus

Segmantal bronchus

Alveoli

Respiratory bronchiole

Cells need oxygen to produce the energy to carry out their roles. Oxygen enters the body through the lungs during breathing. Adult lungs can hold about six litres of air, but every breath moves only about half a litre. The airways that lead into the lungs branch off into smaller bronchial tubes that end in air sacs called alveoli. The alveoli are directly next to blood vessels. Oxygen moves from the alveoli to the blood, where it is picked up by red blood cells. Meanwhile, carbon dioxide moves from the blood to the alveoli, from where it is breathed out.

# 58

# Blood

Blood is a mixture of different types of blood **cells** and plasma. By far the most abundant cell type in blood is the red blood cell. These cells are shaped like discs, and they are the only cells in our body that have no **nucleus**. What they do have is a **protein** called **haemoglobin**, which has the important task of transporting oxygen through the body.

The haemoglobin protein was one of the first proteins of which scientists discovered the three-dimensional protein structure.

Red blood cells have different combinations of sugar groups on the outside. This is what determines the different blood types A, B, AB and O. In addition, different proteins on the red blood cell determine whether the blood is positive or negative for Rhesus (Rh) **antigens**. Knowing someone's blood type is important for blood transfusions, because the blood has antibodies that will attempt to fight off blood cells with different antigens (sugar or protein groups).

Haematopoietic stem cells in bone marrow produce red blood cells, platelets and different types of white blood cells.
White blood cells are part of the immune system, while platelets work together with proteins to create blood clots to heal wounds. Most blood cells are red blood cells, which transport oxygen from the lungs to the rest of the body using the protein haemoglobin. On the outside of red blood cells are different types of sugars and proteins These determine blood types A, B, AB and O as well as the Rhesus (Rh) blood type (often shown as + or −).

Red blood cells, platelets and white blood cells are the three types of blood cells.

Red blood cells

Platelets

White blood cells

# 59

# Heart and circulatory system

The circulatory system is like a busy freight route. Every minute, the heart pumps about five litres of blood through arteries and veins that reach all parts of the body. Thanks to this system, red blood **cells** bring oxygen from the lungs, immune system components fight infections wherever they occur, and **hormones** or other **compounds** in the blood travel to where they need to be. Blood vessels start out big with the aorta near the heart, but branch off into smaller arteries down to tiny capillary vessels in **tissues** throughout the body. From there, the blood makes its way back via veins that get larger as they get closer to the heart.

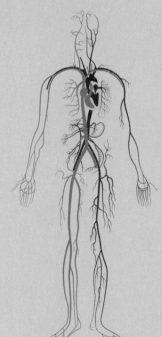

Diagram of the human circulatory system. Arteries leave the heart and veins return blood to the heart. Most arteries are oxygen-rich (shown in red), except for the oxygen-poor (blue) pulmonary artery which goes from the heart to the lungs.

The heart and blood vessels together make up the circulatory system, which pumps blood through the body. Blood transports oxygen (from the lungs), **nutrients**, hormones and other compounds to cells throughout the body. Waste products are also removed via blood.

The heart has four chambers that work together to make sure that blood is continuously pumped in the right direction through the circulatory system. Arteries carry blood from the heart, while veins bring blood back to the heart again. The biggest artery just after the heart is the aorta, but they get smaller as they move further through the body.

Diagram of a cross-section of the heart as seen from the front (so the left atrium and ventricle are shown on the right, and vice versa).

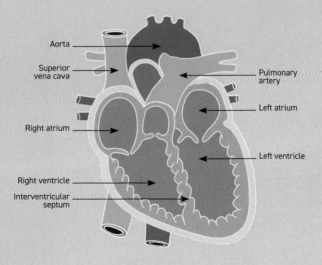

Aorta

Superior vena cava

Pulmonary artery

Left atrium

Right atrium

Left ventricle

Right ventricle

Interventricular septum

# 60

# Hormones

More than 50 **hormones** regulate different processes in the body. For example, adrenaline helps you quickly respond to danger, while melatonin regulates when you fall asleep and wake up.

Many hormones are produced by the endocrine system, which includes organs such as the adrenal glands, pancreas, ovaries and testes.

There are three chemically distinct types of hormones: amine hormones (which resemble an **amino acid**), peptide hormones (a short chain of amino acids) and steroid hormones (which are derived from the lipid cholesterol). Usually, hormone levels will be carefully regulated to maintain balance.

**WHY IT MATTERS**
Changes in hormone levels regulate many processes in the human body, from sleep to sex

**KEY THINKERS**
Insulin:
Frederick Banting (1891–1941) and Charles Best (1899–1978)
Adrenaline: Takamine Jōkichi (1854–1922)
Human growth hormone: Choh Hao Li (1913–1987)
Measuring hormones: Rosalyn Yalow (1921–2011)

**WHAT COMES NEXT**
Several different hormones play a role in growth during childhood and in sexual development during puberty (see Growth and growing up, p.94)

**SEE ALSO**
Cell signalling and transport, p.50
Heart and circulatory system, p.112
Homeostasis, p.130

In
**100**
words

Hormones are messenger molecules that are transported in the blood to cells throughout the body. They are all derived from amino acids, from small peptides or from cholesterol.
Cells receive signals from hormones when they bind to hormone **receptors** either on the surface of the **cell** or inside of it. Hormone levels change in response to what's happening in the body. For example, the pancreas produces more insulin when glucose levels in the blood are high.
There are more than 50 human hormones with many different roles, such as managing sleep–wake cycles, growth and stress response.

## 61

## In 100 words

The lymphatic system is a network of vessels and organs which transports lymph (fluid and white blood cells) through the body.
One of the purposes of the lymphatic system is to pick up excess fluid that leaked from blood vessels to surrounding tissues. The lymphatic system brings this fluid back to the blood, together with waste products such as dead cells. The lymphatic system also produces and activates white blood cells, or lymphocytes. These cells are an important part of the immune system because they create antibodies and remove **pathogens**. Lymph nodes swell when they are active during an infection.

**WHY IT MATTERS**
The lymphatic system is an important part of the immune system (see p.115) because it forms and activates white blood cells

**KEY THINKERS**
Olaus Rudbeck (1630–1702)
Thomas Bartholin (1616–1680)
Alexander Monro (1697–1767)

**WHAT COMES NEXT**
The lymphatic system is the route by which cancer cells can make their way to other parts of the body, so understanding this system is important in preventing and detecting cancer metastasis (see Cancer, p.98)

**SEE ALSO**
Blood, p.110
Heart and circulatory system, p.112
Immune system, p.116
Antibodies, p.165

# Lymphatic system

Alongside the circulatory system, there is another network of vessels in the body: the lymphatic system. One purpose of the lymphatic system is to remove excess fluid and waste from **tissues** and deliver it to the blood. But it's also part of the immune system.

Besides vessels, the lymphatic system includes organs where **lymphocytes** are formed or activated, such as the thymus (which is active during childhood to train the immune system) and spleen (which produces immune **cells** and removes old red blood cells from the body).

# 62

# Immune system

The immune system is a collective term for many different organs and molecules that work together to fight off disease. It includes the lymphatic system, but it also includes several different types of **lymphocytes**, antibodies, cells that destroy other cells, and **proteins** that immune cells use to communicate.

It's a complex and tightly regulated system. First, the innate, or non-specific, immune system acts quickly to remove anything it encounters that doesn't belong in the body. The adaptive, or specific, immune system is a bit slower to respond, but it creates antibodies that recognise **pathogens** and remembers them for the next time a similar infection happens.

Key components of the innate immune system are phagocytes (such as neutrophils or macrophages), natural killer cells and the complement system. The adaptive immune system's main players are two types of white blood cells, T cells and B cells, which together produce specific antibodies against the infectious pathogen.

The immune system protects the body against pathogens. Many different types of **cells** and molecules collaborate to maintain immunity. Phagocytes recognise **antigens** that are foreign to the body and destroy microbes that don't belong. Inflammation and several complement proteins also help this process. Meanwhile, natural killer cells destroy the body's own cells if they are compromised by infections or cancer. Phagocytes display the antigens they attacked on the outside of their cell, from where they bind to T cells and activate those. Some of the T cells further activate B cells, which produce antibodies that more specifically target the infection.

Cells of the immune system include granulocytes (basophils, eosinophils and neutrophils), natural killer (NK) cells, lymphocytes, macrophages, dendritic cells and fibroblasts.

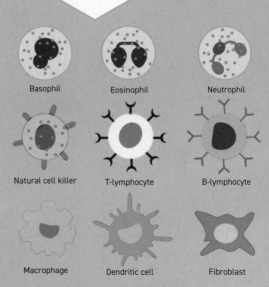

Basophil

Eosinophil

Neutrophil

Natural cell killer

T-lymphocyte

B-lymphocyte

Macrophage

Dendritic cell

Fibroblast

# 63

# Antibodies

**WHY IT MATTERS**
Antibodies recognise
the difference between
antigens on pathogens
and those on healthy
cells in your body. This
is how they fight off
disease.

**KEY THINKERS**
Antibody production in
B cells: Max Cooper
(1933–)
Antibody genetic
diversity:
Susumu Tonegawa
(1939–)

**WHAT COMES NEXT**
Vaccination with
inactive forms of
pathogens or their
antigens activates B
cells to produce
antibodies against this
type of infection (see
Vaccines, p.166).
Because antibodies
recognise specific
targets, they're also
used as tools in
molecular biology
research (see
Measuring with
antibodies, p.150).

**SEE ALSO**
Viruses, p.22
Microbiome, p.74
Immune system, p.116

Antibodies work by binding to **antigens**. In the context of the immune response, an antigen can be, for example, a peptide or a polysaccharide on a bacterial **membrane** or part of a virus envelope.

The immune system ignores antigens from your own cells or those of your microbiome, but when it spots a strange new antigen, it kicks off the immune response. Part of that involves B **cells** making new antibodies specific to this antigen.

There is a lot of variation in the **antibodies** that B cells can make, because antibody **genes** have a lot of smaller segments that can be combined in different ways. In addition, B cells with **genetic variants** that have a stronger interaction with the antigen are more likely to produce more antibodies. Through this system, it's possible for the human body to form antibodies against any antigen and boost the production of exactly the one it needs more of.

Antibodies are Y-shaped **proteins** that recognise a specific antigen on the surface of a **pathogen**. They are produced by B cells. Each B cell makes a different antibody by only expressing certain segments of the antibody gene. B cells that produce the antibody which best matches the antigen will be activated to create more of these antibodies and release them into the blood where they bind to the pathogen.

Some T cells and B cells become memory cells that remember antigens and antibodies after an infection. If the same antigen appears again, antibody production will be faster and more effective.

Antibodies have heavy chains and light chains that together form a Y-shaped protein structure. The variable region of an antibody binds to a specific antigen.

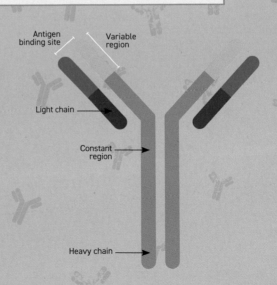

Antigen binding site

Variable region

Light chain

Constant region

Heavy chain

# Skin

The skin is the largest organ in the body, and the most visible.

It's the first layer of defence against any infections, but it also protects the inside of the body from extreme temperatures, UV rays and other dangers. The skin also has touch **receptors** that transfer information about textures or temperatures to the brain.

Skin cells are regularly replaced by new ones that originate from stem cells in the epidermis layer of the skin.

**WHY IT MATTERS**
The skin is the largest organ of the human body and is the first line of defence against infections or damage.

**KEY THINKERS**
Skin anatomy:
Andreas Vesalius
(1514–1564)
Skin development and renewal:
Elaine Fuchs (1950–)

**WHAT COMES NEXT**
Other organs also have epithelial layers, including the intestines and the inside of the mouth (see Ear, nose, mouth, throat, p.121 and Nutrition and digestion, p.126)

**SEE ALSO**
Stem cells and cell differentiation, p.96
Blood, p.110
Immune system, p.116

## In 100 words

The outermost **epithelial tissue** layer of the skin is called the epidermis. The most common **cell** type in this layer are keratinocytes, which are formed from stem cells in the deepest part of the epidermis. Another type of skin cell are melanocytes, which produce the skin pigment melanin. Most people have roughly the same number of melanocytes, but they differ in how much melanin they produce. When skin is damaged, this activates a series of wound-healing events, such as blood clotting to form a scab and the recruitment of white blood cells to fight infections that enter through the wound.

## In 100 words

The ears, nose, mouth and throat form a connected system. They're often studied together, and conditions related to one of these organs may affect the others. However, they all have unique functions as well.

The ear picks up sound waves that reach the ear drum and transfers the signal to three minuscule bones, which connect to the nervous system.

Olfactory **receptors** in the nose and taste buds on the tongue distinguish different molecules to pass on scent or taste signals to the brain. The nose and mouth both connect to the throat, which passes air and food down different routes.

# Ears, nose, mouth, throat

If you've ever relieved pressure from your ears by swallowing, or accidentally sneezed a drink out through your nose, you'll have realised that the ears, nose, mouth and throat are all connected. We think of them as different organs because the experiences of hearing sound and smelling a scent are so different, but from a practical point of view they're all related. There is even a field of medicine, otorhinolaryngology, which specialises in ears, nose and throat issues.

**WHY IT MATTERS**
The ears, nose, mouth and throat together form entry points for information, nutrition and air

**KEY THINKERS**
Inner ear:
Bartolomeo Eustachi
(c. 1500–1574)
Olfactory receptors:
Richard Axel (1946–)
and Linda Buck
(1947–)
Taste buds:
Georg Meissner
(1829–1905) and
Rudolf Wagner
(1805–1864)

**WHAT COMES NEXT**
Sounds, scents and taste pass on signals to the brain to help us make sense of the world (see Brain and nervous system, p.124)

**SEE ALSO**
Lungs and breathing, p.108
Eye, p.122
Nutrition and digestion, p.126

# 66

# Eye

**WHY IT MATTERS**
The eyes are the only
organ that pick up
visual information
from the environment

**KEY THINKERS**
Cones and colour vision:
Thomas Young
(1773–1829)
Retinal in light detection:
George Wald
(1906–1997)

**WHAT COMES NEXT**
Light signals from the
eye are passed on to the
brain via the optic nerve
(see Brain and nervous
system, p.124)

**SEE ALSO**
Cells, p.26
Ear, nose, mouth,
throat, p.121

When you look at something, like a book or a tree, what you observe is actually the light that's being reflected off the object.

Different wavelengths of light are interpreted in your brain as different colours. But for this signal to get to the brain, the information first has to be translated from light patterns to biochemical signals. That's the role of rods and cones in the retina of the eye.

Rod and cone cells detect whether the light intensity is high or low, as well as its wavelength. Thanks to their arrangement throughout the eye, they can also see the direction from which different light signals are coming. Much like pixels on a screen, the signals from millions of rods and cones are able to recreate the image you see.

> **"**On the earth, even in the darkest night, the light never wholly abandons his rule. It is diffused and subtle, but little as may remain, the retina of the eye is sensible of it.**"**
>
> **Jules Verne,** author

Light enters the eye through the pupil, an opening in the iris that can change size to adjust for light levels. After travelling through the lens and vitreous fluid, the light hits rod and cone cells in the retina at the back of the eye.

Cones detect different wavelengths (colours) of light. Rods detect light even at low levels. Both have light **receptors** that contain retinal, derived from vitamin A. Retinal transfers the light signal through the receptor to the rest of the **cell**. Signals from rods and cones are passed on to the nervous system via the optic nerve.

Cross-section of an eye. Light enters the pupil and projects an image on the retina. Rods and cones in the retina pass on the signal to the neurons in the optic nerve.

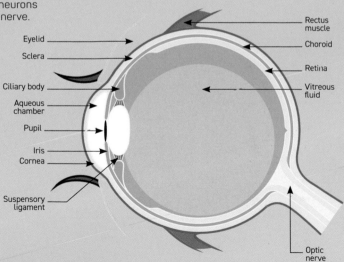

Rectus muscle

Choroid

Retina

Vitreous fluid

Eyelid

Sclera

Ciliary body

Aqueous chamber

Pupil

Iris

Cornea

Suspensory ligament

Optic nerve

# 67 Brain and nervous system

Your brain and nervous system are always busy. Whether you're sleeping, reading, eating or walking, there are always signals that need to pass from one part of the body to another. The brain is the centre where all these signals come together to be processed.

Different regions of the brain each have their own speciality, such as vision, movement, memory, speech, problem solving, spatial perception, hormone production, and temperature regulation.

Cranial nerves reach from the brain to other parts of the head, such as the ears and eyes, while the spinal cord connects **neurons** between the brain and the rest of the body.

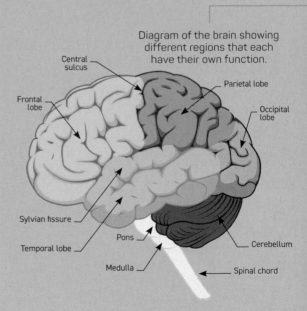

Diagram of the brain showing different regions that each have their own function.

Central sulcus

Frontal lobe

Parietal lobe

Occipital lobe

Sylvian fissure

Pons

Temporal lobe

Medulla

Cerebellum

Spinal chord

Information from the whole of the body and signals received from outside are all processed in the brain. Different parts of the brain specialise in certain functions. For example, the hippocampus specialises in learning and memory. The cells that convey signals within the brain and nervous system are called neurons or nerve **cells**. They interact with each other by transferring neurotransmitter molecules from one neuron to the next. The area where two neurons meet is called a synapse. New connections between neurons can be formed over time, and old ones can be removed. This is called neuroplasticity or brain plasticity.

Neurons communicate by releasing vesicles with neurotransmitters which are recognised by the next cell.

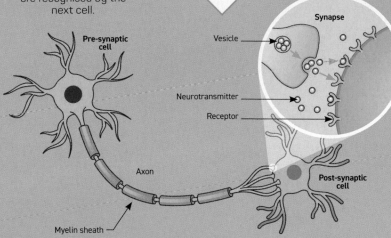

Synapse

Vesicle

Pre-synaptic cell

Neurotransmitter

Receptor

Axon

Post-synaptic cell

Myelin sheath

# 68

# Nutrition and digestion

**WHY IT MATTERS**
The digestive system
turns food into
nutrients that
cells can use

**KEY THINKERS**
Galen (129–216)
William Beaumont
(1795–1853)
Claude Bernard
(1813–1878)

**WHAT COMES NEXT**
Water from food and
drinks is absorbed
from the intestine into
the blood. From there,
blood passes through
the kidneys where any
waste is processed
(see Kidneys and
urinary system, p.128)

Most of the **nutrients** that cells need to function come from food. But to access these nutrients from the food we eat, they need to be broken down into smaller components. Carbohydrates such as starch need to become simple sugars, fats need to become fatty acids, and **proteins** need to be broken down into **amino acids**. That's the role of **enzymes** in the digestive system or gastrointestinal (GI) tract.

Digestive enzyme names usually end in "-ase", and the first part of their name suggests what they break down. For example, lipases turn lipids into fatty acids, proteases cut proteins into amino acids, and lactase breaks down lactose into monosaccharides.

The digestive system starts at the mouth, where the first enzymes are already digesting food, and passes through the oesophagus, stomach, small intestine and large intestine.

Digestion starts in the mouth, where chewing cuts food into smaller chunks while the enzyme amylase breaks down larger carbohydrates into smaller ones. Other enzymes in the stomach and small intestine further digest carbohydrates, fats and proteins into subunits. Digestive enzymes are often produced in the pancreas, but also by bacteria in the intestines. After nutrients are taken up by the intestines, they reach the blood and are transferred to the liver for further processing. The liver also produces bile to make fat digestion easier. The remaining food reaches the large intestine, which absorbs water from food and produces faeces.

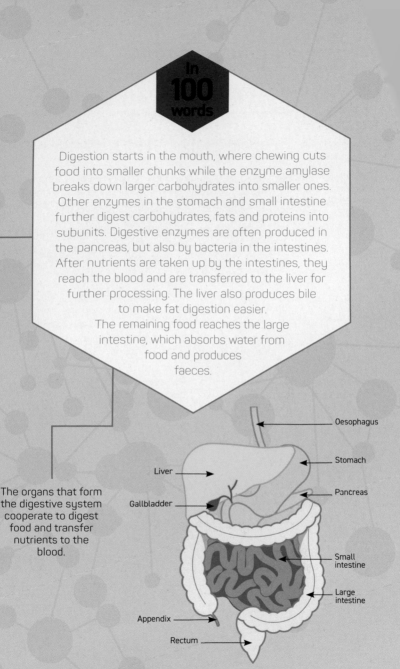

The organs that form the digestive system cooperate to digest food and transfer nutrients to the blood.

Oesophagus

Stomach

Liver

Pancreas

Gallbladder

Small intestine

Large intestine

Appendix

Rectum

# 69 Kidneys and urinary system

Anything you drink eventually turns into urine. But it's quite a journey before it gets there. The water you consume via drinks or food is absorbed from the intestines into the blood. After this, there are still several steps to go before excess fluids are eventually removed from the body.

The kidneys first filter the blood to remove toxins and other waste, while making sure that **nutrients** are still there and that the blood has the right amount of water – not too much and not too little. Any excess water and waste that the kidneys remove from the blood are then passed on to the bladder. From there, urine is removed via the urethra, which is controlled by both involuntary and voluntary muscles.

"If the universe is bigger and stranger than I can imagine, it's best to meet it with an empty bladder."

**John Scalzi** in *Old Man's War*

The kidneys filter waste and excess fluid from blood that arrives from the intestines. Humans have two kidneys, which each have about a million nephrons. These are the functional units of the kidney. Blood vessels first meet the nephron tubule at a place called Bowman's capsule, where water and small molecules are filtered from the blood. Blood vessels stay close to the tubule throughout the nephron, so that they can reabsorb required water and useful components. After the blood is balanced with the right amount of water and nutrients, what's left in the nephron moves to the bladder for excretion.

Kidneys filter the blood and deliver waste to the bladder so that it can be removed.

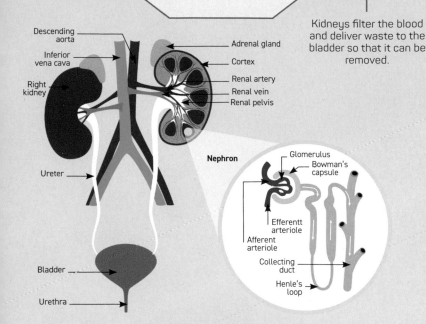

Descending aorta
Inferior vena cava
Right kidney
Ureter
Bladder
Urethra

Adrenal gland
Cortex
Renal artery
Renal vein
Renal pelvis

**Nephron**

Glomerulus
Bowman's capsule
Efferentt arteriole
Afferent arteriole
Collecting duct
Henle's loop

# Homeostasis

**WHY IT MATTERS**
It's all about balance!
Homeostasis keeps the
body's environment
within a range where
its many different
processes can
properly function

**KEY THINKERS**
Claude Bernard
(1813–1878)
Walter Bradford
Cannon (1871–1945)

**WHAT COMES NEXT**
When homeostasis
fails, it can
lead to disease
or death. Depending
on the problem,
organ transplants
(such as a kidney
transplant) might
be a solution (see
Transplants and blood
transfusion, p.168)

**SEE ALSO**
Ageing, p.102
Hormones, p.114
Brain and nervous
system, p.124
Nutrition and
digestion, p.126
Kidneys and urinary
system, p.128

Temperature, blood pressure, blood acidity, glucose levels, hormone levels and other conditions all need to be tightly controlled for cells to do their work. These conditions are kept in check through a series of processes that together are called homeostasis.

Negative feedback loops in homeostasis first detect if a physiological condition is outside of its normal range. They then communicate this to an effector which adjusts the condition back to within its usual levels. For example, the kidneys act as effectors for maintaining blood pressure levels.

Homeostasis is the collective term for many different negative feedback responses that keep the body within a range where it can function properly. For example, the nervous system detects when the body temperature is out of its normal range and starts a negative feedback loop to adjust the temperature through shivering, sweating or other steps. Another example of homeostasis is the pancreas regulating blood glucose levels by balancing the levels of the **hormones** glucagon and insulin. An imbalance in this process leads to diabetes. Many organs are involved in maintaining homeostasis, such as the brain, nervous system, kidneys and pancreas.

# Measuring the Living World

Technology has made it possible to view or detect aspects of biology that were previously invisible. We can measure brain activity, zoom in on cells, calculate the shape of a protein, read genetic code and so much more.

Detection methods and research tools are becoming more advanced and affordable. This means that biologists are collecting more information about life than ever before and now face the challenge of making sense of all of that new information.

# 71

# Ancient and medieval biology

## WHY IT MATTERS
Scientific knowledge is built on previous knowledge. Even if we now know that the old theories were not always the best explanations, they formed a stepping stone for newer discoveries.

## KEY THINKERS
Aristotle
(384–322 BCE)
Maharshi Charaka
(c. 100)
Galen (129–216)
Ibn Sina (980–1047)
Su Song (1020–1101)
Hildegard of Bingen
(1098–1179)

## WHAT COMES NEXT
In the Renaissance period in Europe, the systematic study of natural history and human biology was increasingly popular. This included the use of dissection to study anatomy (see Dissection, p.135).

## SEE ALSO
Taxonomy, p.10
Plants, p.14
Bone and muscle, p.107
Agriculture, p.162
Drug discovery and development, p.167

People have studied the living world for thousands of years to grow food and to stay healthy. Initially, that information was shared by word of mouth, but once people started writing things down, we can see what they learned about biology.

Some cultures started documenting information earlier than others. As a result, we know more about ancient biology from the Middle East, Europe and Asia than from other parts of the world. However, from current indigenous cultures and old artefacts, we know that people everywhere have explored botany, nutrition, human biology and traditional medicine for many centuries.

## In 100 words

For at least 2,000 years, scholars have studied the natural world through observations. They didn't have the tools we have now, but many of their early theories are recognisable as the study of biology. For example, ideas about the role of humours (fluids) of the body emerged in both India and ancient Greece. Meanwhile, Aristotle developed one of the first taxonomies. Medical knowledge from the ancient Greeks and Romans continued to be used in the Middle Ages. During this time, some scholars also wrote about biology, even though it was not yet studied at the early universities of this era.

## 72

Dissection is the systematic study of the internal organs of dead animals or humans. Much of our early medical knowledge came from detailed documentation of dissections. It's still used as a teaching tool. For example, some schools dissect frogs in biology lessons, and medical students might study human cadavers. The human cadavers used in medical training often come from people who donated their body to science. To address ethical concerns about animal welfare or objections to the use of human bodies, it's possible these days to replace dissection in education with technological alternatives such as virtual reality anatomy explorations.

# Dissection

Dissection is one of the oldest ways to study the **anatomy** of animals and humans. Even as early as 3000 BCE, ancient Egyptians learned a lot about the human body from preparing bodies for mummification.

Over the next millennia, people in many different parts of the worlds used dissection to learn about medicine and the human body. During periods when the use of human cadavers was not allowed, early anatomists such as Galen would study animals instead and learn from the similarities between **species**.

**WHY IT MATTERS**
Dissection is one of the earliest experimental methods for learning about human and animal biology.

**KEY THINKERS**
Ancient Egyptians (c.3000 BCE)
Galen (129–216)
Ibn al-Nafis (1213–1288)
Andreas Vesalius (1514–1564)

**WHAT COMES NEXT**
The observations made from dissections became much more detailed after the invention of the light microscope (see p.136).

**SEE ALSO**
Tissues and organs, p.106
Bone and muscle, p.107
Heart and circulatory system, p.112

# 73

# Light microscope

## WHY IT MATTERS
Being able to see individual cells, rather than only larger tissues and organs, opened up entirely new fields of biology such as microbiology, cell biology and biochemistry. Modern medicine relies on research from these fields.

## KEY THINKERS
Antonie van Leeuwenhoek (1632–1723)
Robert Hooke (1635–1703)
Carl Zeiss (1816–1888)
Marvin Minsky (1927–2016)

## WHAT COMES NEXT
Some things are too small even for a light microscope, but with the invention of electron microscopy even more detail became visible (see Electron microscope, p.144).

## SEE ALSO
Cell culture, p.143
Measuring with antibodies, p.150
Fluorescent proteins, p.151

Until the 17th century, people who studied **anatomy** or natural history were limited to what they could see with their own eyes. Anything smaller than a hair or an insect was incredibly difficult to see. Nobody had ever seen bacteria, even though people suspected **germs** existed. Nobody knew that organs and **tissues** were made up of smaller cells. That all changed with the invention of the light microscope. It made the invisible visible.

Many newer microscopes are directly connected to a computer so that the images appear on screen. Until a few decades ago, this was not possible, and it was common to draw the images that you saw through the eyepiece.

Fluorescence microscopes detect **fluorescent** light of specific wavelengths. This allows **cell** biologists to view cells that express **genes** with fluorescent tags or that interact with fluorescent antibodies (see Fluorescent proteins, p.151 and Measuring with antibodies, p.150).

The light microscope directs a source of light through a thin **sample**, such as a slice of a plant or tissue. By looking at the sample from the other direction through an objective lens attached to an eyepiece, the sample can be magnified up to about two thousand times.

Nowadays, cell biologists often use a fluorescent microscope, which detects fluorescent labels in cells. A variation of this is confocal microscopy, which shines light only on a very thin cross-section of the sample so that there is no background signal from the rest of the sample. This creates very sharp images.

Light microscopes use lenses to magnify a specimen illuminated by a light source. Many microscopes can easily switch between several objective lenses to change the level of magnification.

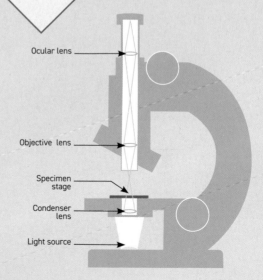

Ocular lens

Objective lens

Specimen stage

Condenser lens

Light source

137

# Statistics

**WHY IT MATTERS**
In biology, no two
samples are identical.
Statistics helps
biologists to make
sense of this variation.

**KEY THINKERS**
Distribution and
probability:
Thomas Bayes
(1702–1761)
Correlation:
Francis Galton
(1822–1911)
Visualising statistics:
Florence Nightingale
(1820–1910)

**WHAT COMES NEXT**
Drug development is
one of the areas where
biostatisticians play an
important role, because
they can determine
whether a new
medication really
works or whether the
results they saw were
just a coincidence (see
Drug discovery and
development, p.167).

**SEE ALSO**
Populations and
ecosystems, p.62
Bioinformatics, p.158

The average height of people who lived 200 years ago is less than the average height of people today. But even today, you will find people who are shorter than the average person of the 1820s, and there were people back then who were taller than the average height of today. This is just one example of the wide variation that occurs in any biological **trait**.

This variety can make it difficult to study biology. How do you know if your observations of a few samples are an accurate representation of an entire **population**?

That's why statistics is important for biologists. Statistics brings fixed mathematical rules to dealing with the many variables that occur in the living world.

"Statistics is the grammar
of science."
**Karl Pearson,** mathematician

In biology, statistics is used to determine whether an observed trait is part of a normal distribution of variable traits or whether it's more likely to be an unusual trait. When a measurement is statistically significant, it's probably not the result of random chance.

Statistics can also determine correlation between properties. For example, animals that eat more food might also give birth to more young. A correlation doesn't show a causal link, but it's a starting point for further research.

It can be difficult to see patterns in the numbers, but graphs or other visualisations can help them stand out.

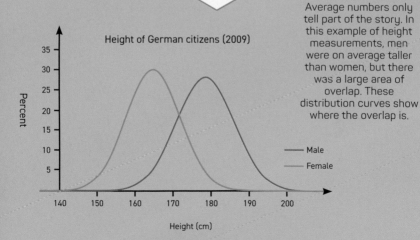

Height of German citizens (2009)

Percent

Height (cm)

— Male
— Female

Average numbers only tell part of the story. In this example of height measurements, men were on average taller than women, but there was a large area of overlap. These distribution curves show where the overlap is.

# 5 Measuring heart and circulation

Electrocardiography (ECG) and non-invasive blood pressure monitoring have been around since the 19th century. These methods made it much easier to monitor people's heart and circulatory health, because before this it wasn't possible to measure heart activity or blood pressure without having to cut into the body.

Electrocardiography detects electric activity from the heart to create a graph (electrocardiogram) that shows when different muscles in the heart are active. The waves and peaks in the graph are indicated by the letters P, Q, R, S and T. Measuring these waves and the distance between them is an indication of what is happening inside the heart.

Blood pressure measurements detect pressure from the main artery in the arm. If blood pressure is outside of the usual range, this can indicate a problem with the heart or kidneys. High blood pressure also increases the risk of developing health conditions.

Newer methods to look in more detail at the heart and circulatory system include ultrasound (using sound waves), PET **scan** (using slightly radioactive material) or MRI imaging (using magnetic fields).

## In 100 words

Electrical pulses from the nervous system signal to the heart muscles when they need to contract. Electrocardiography (ECG) measures these signals through electrodes on the skin and produces a graph that shows whether the heartbeat follows the usual pattern of heart muscle contractions. Blood pressure can also be measured from outside the body using a device that measures pressure in the artery of the arm. Systolic pressure is the highest pressure during a heartbeat, and diastolic pressure is the lowest pressure between heartbeats. Other methods to look at the heart and blood flow include ultrasound, PET scan and MRI imaging.

Waves and peaks in an electrocardiogram show electric activity from the heart and are an indication of heart health.

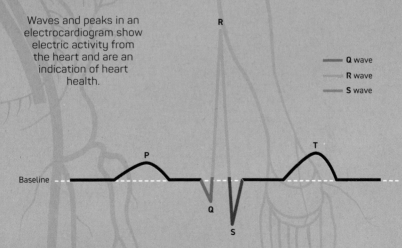

— **Q** wave
— **R** wave
— **S** wave

R

P

T

Baseline

Q

S

# 76

# Measuring the brain

It's incredibly difficult to understand exactly how our brains and thoughts work. However, there are ways to get some idea of what's going on in there. Three technologies that measure certain aspects of brain shape or brain activity are EEG, MRI and fMRI.

These methods can be used to diagnose brain conditions such as epilepsy or brain tumours. They're also used in research to learn more about how the brain functions. For example, MRI **scans** of taxi drivers who memorised all the streets in London showed that their hippocampus (a memory area in the brain) got bigger.

**WHY IT MATTERS**
Measuring brain activity or shape can detect neurological problems or brain health issues. It's also used in research to learn more about how the brain works.

**KEY THINKERS**
EEG: Hans Berger (1873–1941)
MRI: Raymond Damadian (1936–2022)
fMRI: Seiji Ogawa (1934–)

**WHAT COMES NEXT**
These brain monitoring methods can't see what's happening inside individual neurons. To study processes inside a cell, researchers often use cell culture methods combined with microscopy (see Cell culture, opposite).

**SEE ALSO**
Brain and nervous system, p.124
Measuring heart and circulation, p.140

## In 100 words

Three popular methods to measure the brain are electroencephalography (EEG), magnetic resonance imaging (MRI) and functional MRI (fMRI).
EEG works through electrodes placed on the head. These measure electrical activity of **neurons** in different parts of the brain. This is used to diagnose epilepsy, for example. MRI creates a detailed image of the brain that can be used to locate tumours or see other unusual structures in the brain. fMRI shows (on an MRI image) which areas of the brain receive more blood flow, which indicates activity. This can be used to show how the brain responds to certain signals.

Cell or molecular biologists grow cells in dishes or flasks to study what happens inside the cell under certain conditions. For example, by genetically expressing different forms of a protein, researchers can study how each variant responds to signals in the cell. These methods are used to find out the function of each **protein** in a cell or why certain **genetic** mutations lead to disease.
Cell culture uses either primary cells, which are isolated from **tissues** and organs, or immortalised cell lines, which have been kept in culture for many years and are shared by many researchers around the world.

# Cell culture

In 1951, a woman called Henrietta Lacks died of cervical cancer. Researchers noticed that her cancer **cells** stayed alive and kept dividing when they were grown in dishes in the lab. Even decades later, these HeLa cells are regularly used in research — for example, in studies to test new drugs.

There are many other cell lines that have been used in this way since the early 20th century. Cell culture makes it possible to study biology at a cellular level without having to use entire animals or humans.

**WHY IT MATTERS**
By culturing cells that stay alive outside of the body, it's possible to see how human and animal cells respond to different conditions, treatments, or genetic changes.

**KEY THINKERS**
Wilhelm Roux (1850–1924)
Ross Granville Harrison (1870–1959)
George Gey (1899–1970)

**WHAT COMES NEXT**
To label cells or proteins in cell culture, researchers often use fluorescent proteins or antibodies (see Fluorescent proteins, p.151 and Measuring with antibodies, p.150).

**SEE ALSO**
Cells, p.26
Gene expression, p.35
Light microscope, p.136

# Electron microscope

**WHY IT MATTERS**
Electron microscopes can see much smaller details than a light microscope, including organelles within a cell, viruses and **proteins**.

**KEY THINKERS**
TEM: Ernst Ruska (1906–1988), Max Knoll (1897–1969)
SEM: Manfred von Ardenne (1907–1997)
CryoEM: Joachim Frank (1940–), Jacques Dubochet (1942–) and Richard Henderson (1945–)

**WHAT COMES NEXT**
Although cryoEM is becoming a popular way to look at proteins, most protein structures so far have been discovered using X-ray crystallography (see p.146).

**SEE ALSO**
Viruses, p.22
Protein structure, p.40
Mitochondria and chloroplasts, p.53
Light microscope, p.136

Electron microscopes have made even more of the microscopic world visible than light microscopes ever could. As the name suggests, these machines use electrons instead of light to study a **sample**, and that reveals much more detail. Even viruses can be seen with electron microscopy.

The samples for electron microscopy need to be very carefully prepared because even the slightest movement or wiggle would ruin the image. There are different ways to fix the sample so that it doesn't move. In cryogenic electron microscopy (cryoEM) samples are frozen very quickly to preserve their shape and keep them still during the **scan**.

Electron microscopes are bigger and more expensive than light microscopes, so not every biologist will have access to them at their lab, but they might have shared use of a machine at their research institute or at a location nearby.

## In
## 100
### words

Electron microscopes magnify samples with electrons instead of light. Two common types of electron microscopes are scanning electron microscopes (SEM) and transmission electron microscopes (TEM). In SEM, the pattern by which electrons bounce off a sample provides information about the structure of the sample's surface. For example, SEM can visualise details of insects' **compound** eyes. With TEM, electron beams pass through a very thin sample to reveal the details within. TEM can be used to show the structure of **organelles** within a **cell**, such as mitochondria. CryoEM is a type of TEM that can even see the structures of proteins.

Scanning electron microscope (SEM) image showing details of an insect's compound eye.

145

# X-ray crystallography

The double helix structure of **DNA** was discovered thanks to X-ray crystallography images taken by Rosalind Franklin. In this method a **sample** is exposed to X-ray radiation. The X-rays scatter off the sample in a pattern that holds clues about its structure.

X-ray crystallography also made it possible to calculate the shape of many **proteins**, starting with **haemoglobin** and myoglobin. Thousands of protein structures have been solved with X-ray crystallography over the last few decades. Other methods to find the structure of a protein include cryoEM and nuclear magnetic resonance (NMR) spectroscopy, which is similar to MRI (see Measuring the brain, p.142).

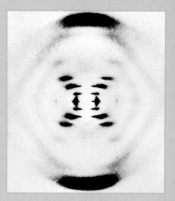

Photo 51 was taken by Rosalind Franklin and her graduate student Raymond Gosling using X-ray crystallography. It was proof that DNA has the shape of a double helix, because only that shape could have created the pattern in this image.

X-ray crystallography can reveal the structure of large molecules, such as DNA and proteins. Thousands of protein structures have been discovered using this method. To carry out X-ray crystallography, the sample first needs to be crystallised, which creates an organised pattern of many copies of the protein. When the **crystal** is exposed to X-rays, the X-rays scatter off the **atoms** in the crystal in a pattern determined by the shape of the molecule. By capturing that pattern, it's possible to calculate the protein's three-dimensional structure, which shows where each **amino acid** is in relation to the rest of the protein.

"I should not like to leave an impression that all structural problems can be settled by X-ray analysis or that all crystal structures are easy to solve. I seem to have spent much more of my life not solving structures than solving them."

**Dorothy Hodgkin,** chemist

# Gel electrophoresis

The difference between two pieces of **DNA** is in the **sequence** of nucleotides. Likewise, two **proteins** are functionally distinct based on the order of their **amino acids**. It's not easy to figure out that order, but there is a shortcut that can quickly compare pieces of DNA and proteins of different size.

Gel electrophoresis sorts DNA or proteins by size. This can be used to detect proteins of a known size in a **sample**, or to isolate a piece of DNA of a certain length. It works by loading samples on a **gel** and applying an electric current that moves the molecules through the gel at different speeds depending on their size.

Once the DNA or proteins are separated by size, they form bands that can be made visible on the gel with stains or **fluorescent** dyes.

It's also possible to transfer ("blot") the bands from the gel to a **membrane** and label them with radioactive tags or antibodies. If this method is used to detect DNA it's called Southern blot, named after its inventor Edwin Southern. Similar techniques to detect **RNA** and proteins were later named Northern blot and Western blot, respectively.

Agarose gels separate fragments of DNA or RNA by size. These gels are connected to an electrical current. DNA or RNA is added to one side of the gel, and the current moves them to the other side. **Restriction enzymes** cut DNA into segments of different size that move through the gel at different speeds. This makes it possible to isolate a specific fragment, for example to create **gene** expression **vectors**.

SDS-PAGE uses acrylamide gels to separate proteins by size. In combination with Western blot, this can be used to confirm the presence of a specific protein in a sample.

Gel electrophoresis can show which size the proteins in a sample are. In this example, both sample A and B include a protein of size 40 kDa (a measure of molecular weight) but sample A also has two other proteins.

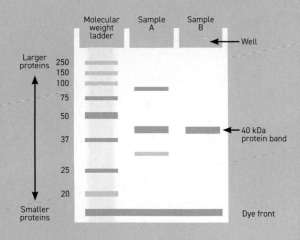

149

# 81

# Measuring with antibodies

Antibodies are not only useful in the immune system. Because they are so good at spotting unique **antigens**, they are also used to label **proteins** in research labs.

Some of the techniques that use antibodies are Western blot (see Gel electrophoresis, p.148), enzyme-linked immunosorbent assay (ELISA) and immunofluorescence (see Light microscope, p.136).

Until recently, antibodies could only be collected from the blood of animals that were exposed to the antigen that the **antibody** needs to detect. However, newer technologies can also create antibodies for research without using animals.

**WHY IT MATTERS**
Antibodies can be used to detect proteins in cells or samples, so that the function of these proteins (and their corresponding genes) can be studied.

**KEY THINKERS**
Immunofluorescence:
Albert Coons
(1912–1978)
ELISA: Eva Engvall
(1940–) and
Peter Perlman
(1919–2005)
Western blot:
Neal Burnette (1944–)

**WHAT COMES NEXT**
Adding fluorescent molecules to antibodies is one way to visualise proteins under a fluorescence microscope. In living cells, another way is to express fluorescent proteins (see Fluorescent proteins, p.151).

**SEE ALSO**
Antibodies, p.118
Light microscope, p.136
Gel electrophoresis, p.148

In **100** words

Antibodies bind to a very specific antigen. In research experiments, this is used to detect proteins — for example, to study their function.
In Western blot, antibodies detect proteins that were separated by size with gel electrophoresis and then transferred to a **membrane**. Each antibody binds to just one protein, so it shows one line on a blot. ELISA uses antibodies to measure the amount of a specific protein or other antigen in a **sample**. It can be used, for example, to diagnose infection by detecting certain antigens in blood. Immunohistochemistry and immunofluorescence use antibodies to label proteins in microscope samples.

Green fluorescent protein (GFP) is used as a reporter gene to test **genetic** editing or to see where a protein is in a cell. The small GFP gene can be added to a plant, animal or cell culture. It can be detected with fluorescence microscopy to see whether cells have successfully taken up **DNA** fragments (including the GFP gene).

GFP can be connected to another protein. The fluorescent signal then reports where in the **tissue** or cell the protein is located.

Other fluorescent proteins in different colours also exist. They are activated using different colour lasers of a fluorescence microscope.

# Fluorescent proteins

The jellyfish *Aequorea victoria*, which lives in the Pacific Ocean, lights up in the dark. It produces its own blue light, which in turn switches on a green fluorescent protein (GFP).

Biologists have been able to take that GFP **gene** and express it in cells from plants or other animals without disturbing the cells. Shining a blue laser light on these cells then shows exactly which cells have the GFP. The main reasons to do this are to visualise another protein in a living **cell** or to test whether a **genetic** manipulation method is working correctly.

**WHY IT MATTERS**
In studies of the function of genes and proteins, fluorescent proteins have been a helpful tool to show what's happening inside a cell.

**KEY THINKERS**
Roger Tsien
(1952–2016)
Osamu Shimomura
(1928–2018)
Martin Chalfie (1947–)

**WHAT COMES NEXT**
GFP is often used in cell culture experiments or in model organisms such as fruit flies to test whether specific genes can be switched on or off on demand (see Manipulating gene expression, p.152).

**SEE ALSO**
Invertebrates, p.16
Gene expression, p.35
Light microscope, p.136

# 83 Manipulating gene expression

**WHY IT MATTERS**
Changing the level of gene expression in cell cultures or model organisms can reveal the functions of specific genes or proteins.

**KEY THINKERS**
Recombinant DNA: Paul Berg (1926–2023)
RNA interference: Andrew Fire (1959–) and Craig Mello (1960–)
CRISPR-Cas9: Jennifer Doudna (1964–) and Emmanuelle Charpentier (1968–)

**WHAT COMES NEXT**
Certain gene editing technologies are being explored as possible future medical treatments of genetic conditions (see Gene therapy, p.180).

**SEE ALSO**
Model organisms, p.20
Cell culture, p.143
Gene expression, p.35
Mutation and variation, p.42
PCR, p.154

A lot of what biologists have learned over the years about the function of different **genes** and proteins has come from **cell** culture and model **organism** studies in which genes or **proteins** were altered or switched on or off to observe what happened. For example, flies with missing Hox genes grow body parts in a different place (see Genes in development, p.90).

These studies often use **recombinant** DNA technology, in which a fragment of **DNA** can be moved from one organism to another. This usually involves a small circular piece of DNA, called a plasmid or **vector**. The vector includes DNA **code** for a gene and its promoter so that the gene can be expressed in a cell that takes up the vector.

DNA can be edited with **restriction enzymes** and ligases. Restriction enzymes cut DNA at specific restriction sites (unique to each enzyme) and ligases paste it together. Modern **gene editing** methods such as CRISPR-Cas9 are more precise, because they can cut anywhere, not just at a restriction site.

Recombinant DNA methods also make it possible to express **fluorescent** proteins (see p.151) or to use bacteria to produce large amounts of a human protein for X-ray crystallography (see p.146).

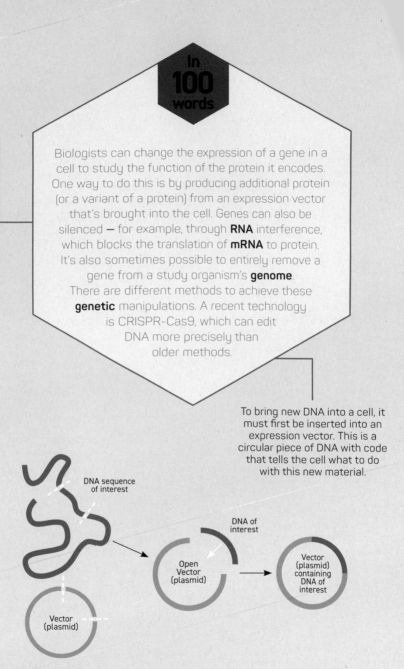

Biologists can change the expression of a gene in a cell to study the function of the protein it encodes. One way to do this is by producing additional protein (or a variant of a protein) from an expression vector that's brought into the cell. Genes can also be silenced — for example, through **RNA** interference, which blocks the translation of **mRNA** to protein. It's also sometimes possible to entirely remove a gene from a study organism's **genome**. There are different methods to achieve these **genetic** manipulations. A recent technology is CRISPR-Cas9, which can edit DNA more precisely than older methods.

To bring new DNA into a cell, it must first be inserted into an expression vector. This is a circular piece of DNA with code that tells the cell what to do with this new material.

DNA sequence
of interest

DNA of
interest

Open
Vector
(plasmid)

Vector
(plasmid)
containing
DNA of
interest

Vector
(plasmid)

# 84

# PCR

## WHY IT MATTERS
Forensics, medical diagnostics, drug development, environmental research, genetics research, and many other modern fields of biology all regularly use PCR to study and manipulate DNA for many different purposes.

## KEY THINKERS
Primer replication:
Har Ghobind Khorana
(1922–2011)
Thermus aquaticus
discovery:
Thomas Brock
(1926–2021)
PCR: Kary Mullis
(1944–2019),
Henry Ehrlich (1943–)

## WHAT COMES NEXT
One of the applications that uses PCR is genetic fingerprinting (see DNA fingerprinting, p.169).

## SEE ALSO
Prokaryotes, p.12
DNA, p.30
Manipulating gene expression, p.152

All cells in all **organisms** use the enzyme DNA **polymerase** to form more **DNA** when needed — for example, during **cell division**. DNA polymerase is also used to create copies of a piece of DNA in a lab. A method called PCR (polymerase chain reaction) does this very rapidly by heating the DNA to pull its double **strands** apart and then using one strand as a template to make a new copy of the second strand.

Most organisms' DNA polymerase **enzymes** could not withstand the heat used in a PCR reaction, so the method uses the DNA polymerase from bacteria that live in hot springs. This is called Taq polymerase, after the microbe's full name *Thermus aquaticus*.

PCR is a very popular method used in almost all molecular biology labs to create **vectors** for **recombinant gene** expression (see Manipulating gene expression, p.152). It's also used in forensics, to amplify tiny amounts of DNA found at a crime scene, and it can detect infectious diseases (such as COVID-19) by looking for the DNA of the virus or bacteria that causes it.

Each PCR cycle is driven by temperature changes that determine when the DNA is pulled apart, when primers bind and when Taq polymerase forms a new strand.

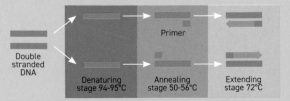

Double stranded DNA

Primer

Denaturing stage 94-95°C

Annealing stage 50-56°C

Extending stage 72°C

The polymerase chain reaction (PCR) method makes many copies of a small section of DNA. It's used to make recombinant DNA vectors, to diagnose an infection from viral or bacterial DNA, or in forensics research.
PCR works by first heating DNA so that the double strand loosens into single strands. The single strands then bind to DNA primers – very short bits of DNA that recognise part of the larger strands. From here, Taq polymerase duplicates the single strand into a double strand again. Every time this cycle is repeated, it copies only the fragment of DNA between the two primers.

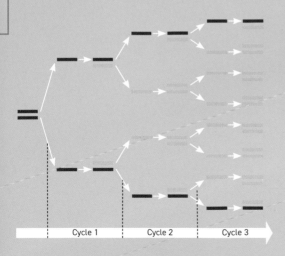

During every PCR cycle, the amount of target DNA is doubled. After three cycles, there is eight times as much DNA. After 30 cycles, it's over a billion times as much as at the start.

Cycle 1    Cycle 2    Cycle 3

155

# 85

# Genetic sequencing

The **sequence** of A, C, T and G in a **gene** holds a lot of information about the function of that gene. But for many years, finding that sequence was not a simple task.

For several decades since the late 1970s, the main method for finding the order of nucleotides in **DNA** was Sanger sequencing. It's similar to PCR in that it uses DNA **polymerase** and nucleotides to grow a complementary **strand** onto single-strand DNA. The difference is that Sanger sequencing reads each new **nucleotide** as it is added. This takes a long time because it only works on short segments of DNA at a time.

A much faster method is next-generation sequencing (NGS), in which multiple fragments of DNA are sequenced at the same time. This has made it easier to sequence entire genomes and to better understand our own and other **species'** genetic material. Whole **genome** sequencing is also increasingly used in medical research, for example to find the genetic causes of rare diseases.

The newest (third generation) sequencing method is nanopore sequencing, which reads an electrical current while a piece of DNA or **RNA** moves through a very small opening, or nanopore.

The genetic **code** reveals information about gene function, evolution or genetic conditions. Sanger sequencing reads a DNA sequence while copying a new strand. Labelled nucleotides at the end of each new segment are detected to decipher the code. This method was used to find the sequence of all genes of the human genome. It took years. Newer and faster methods include NGS, which sequences multiple pieces of DNA at a time, and nanopore sequencing, which measures an electric current to sequence **genetic** material. Small nanopore devices can easily be taken to remote locations to study infectious diseases or monitor biodiversity.

"The cost of sequencing a genome is coming down to of the order of a hundred dollars. It's approaching the cost of many simple diagnostic tests."
**Shankar Balasubramanian,** chemist

# Bioinformatics

Thanks to technology, biologists are able to collect a lot of detailed information, such as **genome sequences**, automated measurements, microscope images or **protein** structures. It's too much information to analyse by hand, but bioinformaticians can make sense of it. They use software to analyse biological data, particularly from experiments that generate so much information that it would be near impossible otherwise. This includes assembling whole genomes, analysing protein structures, detecting new information from large sets of microscope images, discovering new **genetic variants**, predicting how molecules and proteins will interact, and more.

Artificial intelligence and machine learning are increasingly being used for some of these processes. This shows that bioinformatics forms a bridge between biology and new **developments** in computer science.

> "All these molecular technologies are really going to overrun the medical and biodiversity fields and environmental fields because they are so powerful. But they generate lots of data that has got to be looked after!"
>
> **Janet Thornton,** bioinformatics expert

Bioinformatics uses computer science to study large amounts of biological data, such as **genetic** sequences, protein structures, microscope images, or data from automated experiments that study many samples at a time.

There are international standards for the way that genetic sequences or protein structures are shared online with other researchers. This way, the data can be compared and analysed with specialised software designed by bioinformaticians. Bioinformaticians also work directly with laboratory biologists on large automated studies. They can quickly find patterns in datasets and make new observations that can then be tested by researchers doing further studies in the lab.

# Applying Biology

In just the last few decades, biologists have learned how to use DNA to solve crimes, how to teach the immune system to attack cancer cells and so much more.

Sometimes, making changes to the living world can raise ethical issues. Should we bring back extinct animals? Should we edit babies' DNA before birth? Biology can't answer those questions. It can only explain why it is technically possible to do so.

Other applications are less controversial. Most people would agree that we need new types of antibiotics or that we need to protect biodiversity, for example. Applying knowledge from biology can find solutions for those and many other problems.

# Agriculture

Bananas almost went extinct. Because banana farmers, like many other farmers, preferred one particular **genetic variant** of their crop, there was not enough **genetic diversity** to protect the fruit from disease, and a fungus outbreak in the 1950s almost killed off all bananas in the world.

This shows how understanding biology – in this case, genetic diversity – supports agriculture. Another example is the biology of nutrient cycles. Farmers who always grow the same plants in the same location risk depleting nutrients from the soil, so they rotate their crops between locations.

## WHY IT MATTERS
Agriculture supports the world's requirement for food, but it also affects biodiversity and genetic variety. Meeting global food needs in a sustainable way is one of the current challenges in biology.

## KEY THINKERS
Ibn al-'Awwam
(c. 1100–1158)
George Washington Carver (1864–1943)
Norman Borlaug
(1914–2009)
M. S. Swaminathan
(1925–)
Evangelina Villegas
(1924–2017)
Maria Andrade (1958–)
Li Jiayang (1956–)

## WHAT COMES NEXT
Agriculture researchers test that crops are genetically diverse with methods such as genetic sequencing (see p.156) or DNA fingerprinting (see p.169).

## SEE ALSO
Nutrient cycles, p.64
Animal behaviour, p.69
Plant development, p.82

In
**100**
words

One of the earliest applications of biology was in agriculture. For example, knowledge about nutrient cycles encourages farmers to rotate crops to keep soil healthy. Selecting for preferred crops and breeds gradually changed the genetics of cultivated plants and domesticated animals. Genetic engineering speeds up this process. It can be used to increase crop yield by making plants resistant to weed killer or disease. However, plants are at risk of disease if they are genetically identical. Some people also worry that genetically modified organisms (GMOs) will cross with wild plants or bring unknown health effects, so GMOs are carefully regulated.

Conservation biology studies the effect that humans have on different **ecosystems** and prevents the loss of biodiversity. This includes protecting endangered species by monitoring their numbers and setting up breeding programmes with wildlife centres or zoos. **Population** studies, animal behaviour and genetics research discover which animals are at risk of extinction and how to protect them. Conservation biologists also encourage farmers and gardeners to change their habits to protect ecosystems. For example, mowing the lawn less often spurs wildflower growth and attracts pollinators. In some areas, rewilding brings ecosystems back to the state they were before agriculture or timber trade.

**WHY IT MATTERS**
Conservation maintains and restores biodiversity, which makes ecosystems more resilient to disease or climate change.

**KEY THINKERS**
Rachel Carson (1907–1964)
Michael Soulé (1936–2020)
E. O. Wilson (1929–2021)
Idelisa Bonnelly (1931–2022)

# Conservation

In the 1980s, there were only about a thousand giant pandas left in the wild. Recently, that number doubled to almost two thousand thanks to decades of conservation efforts.

Many other **species** on land and in the sea are also actively monitored by conservation biologists who are concerned about their dwindling numbers. Hunting and whaling has directly endangered some animals, but agriculture, industrialisation and the timber trade have also threatened animals and plants by destroying their **habitats**.

Extinction of any species is a threat to overall biodiversity, so it's something that conservation researchers try to prevent.

**WHAT COMES NEXT**
Restoring habitats or rebalancing populations of endangered animals changes food webs and biomass in the ecosystem (see Food webs and biomass, p.66).

# 89

# Germ theory

Even though scientists knew that bacteria existed since they first observed them in the 17th century, it took until the middle of the 19th century for researchers to realise that some bacteria cause decay or disease. Initially they thought that bacteria were attracted to infections or to rotting food, which was why they were often found there. The discovery that bacteria were causing it had major implications across healthcare, science and even regular household cleaning.

**WHY IT MATTERS**
Germ theory encouraged doctors to make sure that their tools and hands were clean before surgery, which saved many patients from unnecessary infections.

**KEY THINKERS**
Disinfection in medical clinics:
Ignaz Semmelweis (1818–1865) and Joseph Lister (1827–1912)
Pasteurisation:
Louis Pasteur (1822–1895)
Different germs cause different diseases:
Robert Koch (1843–1910)

**WHAT COMES NEXT**
Once people understood that microbes could cause disease, they started to look for ways to get rid of them (see Antibiotics, opposite).

**SEE ALSO**
Prokaryotes, p.12
Immune system, p.116
Light microscope, p.136

## In 100 words

In the mid-19th century, Ignaz Semmelweis and Joseph Lister both discovered that the number of patient infections or deaths went down if doctors thoroughly cleaned their hands and tools. Around the same time, Louis Pasteur found a way to prevent wine from going bad by rapidly heating and cooling it, now known as pasteurisation.
Both discoveries pointed in the same direction: if killing **germs** stops food decay and prevents infections, then the germs are causing the decay and disease. This realisation of the germ theory of disease has saved many lives and paved the way for the discovery of antibiotics.

In **100** words

## 90

Antibiotics kill bacteria or stop them from multiplying. Each antibiotic works in a different way. For example, penicillin prevents bacteria from forming their **cell** wall, while other antibiotics might interfere with **enzymes** in the cell. Because antibiotics only kill bacteria, they do not help against viral infections. Not all bacteria are equally sensitive to antibiotics. If a few bacteria that are more resistant to it survive treatment, they can multiply and cause new infections that are more difficult to treat. Antibiotic resistance is a growing global health issue because it can make some diseases untreatable unless new antibiotics are discovered.

**WHY IT MATTERS**
Antibiotics are used to treat bacterial infections such as pneumonia, cholera, syphilis and tuberculosis.

**KEY THINKERS**
Ancient Egyptians (c. 1550 BCE) used moulds to treat infections.
Alexander Fleming (1881–1955) discovered that mould produces penicillin.

**WHAT COMES NEXT**
Finding new antibiotics is important to treat bacterial infections that have become resistant to existing antibiotics (see Drug discovery and development, p.167).

**SEE ALSO**
Germ theory, opposite
Prokaryotes, p.12
Fungi, p.15
Enzymes, p.48

# Antibiotics

When Alexander Fleming returned from holiday in 1928, he discovered mould growing on one of his agar plates with bacteria. He noticed that no bacteria grew close to the mould, which made him realise that there was a substance in mould that could kill bacteria. Fleming called it "penicillin", and it became the first antibiotic to treat bacterial infections.

Several other antibiotics have also been discovered in fungi or in soil microorganisms, which use these **compounds** as a natural defence against bacteria. However, more antibiotics are needed because bacteria are becoming resistant to the existing ones.

# 91

# Vaccines

Before vaccines, the only way to become immune to an infection was to encounter the **pathogen**. For example, rubbing smallpox scabs on a healthy person's scratched skin protected them from a more severe form of smallpox. This method, called variolation, was first used in Asia, the Middle East and Africa.

In 1796, Edward Jenner proved that purposely infecting people with cowpox, a much milder disease, also protected them against smallpox. By training the immune system to recognise a disease without ever seeing that pathogen, Jenner created the first vaccine. The word "vaccine" comes from *variolae vaccinae*, Jenner's term for cowpox.

**WHY IT MATTERS**
Vaccines save many lives. Many thousands of people died of polio before vaccination started in the 1950s. Now polio is almost entirely eradicated worldwide.

**KEY THINKERS**
Variolation: Used in Asia, Africa and the Middle East since the 16th century.
Smallpox vaccine: Edward Jenner (1749–1823)
Polio vaccine: Jonas Salk (1914–1995)
mRNA vaccines: Katalin Karikó (1955–) and Drew Weissman (1959–)

**WHAT COMES NEXT**
Vaccine testing is a rigorous process that takes years (see Drug discovery and development, opposite). However, it can be sped up in emergencies with additional funding and trial volunteers.

**SEE ALSO**
Immune system, p.116
Antibodies, p.158

## In 100 words

A vaccine trains the immune system to recognise a virus or disease-causing bacteria without needing to be infected with this pathogen.
Vaccines include either an inactive form of the pathogen or one of the components of its outer surface to encourage cells of the immune system to create **antibodies** against this **antigen**. This allows the immune system to fight off any virus or bacteria that carries the same antigen. Some vaccines use a small piece of **mRNA** to express the antigen.
While some vaccines last a lifetime, others require regular boosters that are updated for new versions of the pathogen.

In **100** words

Many medicinal drugs target **protein** receptors on the surface of cells. This includes a group of G-protein coupled **receptors**, which naturally interact with components outside of the cell and convey signals within. Drugs can interfere with these pathways. For example, antihistamines prevent histamine from causing allergic reactions by binding to histamine receptors. Discovering **molecules** that affect cell signalling by binding to receptors is now a common way to develop new drugs. This process uses bioinformatics, cell cultures and high throughput screening methods to find drug candidates that are then studied further in a long series of tests and clinical trials.

**WHY IT MATTERS**
Drugs that were either discovered in nature or designed in research labs can treat a wide variety of diseases.

**KEY THINKERS**
Malaria treatment:
Tu Youyou (1930–)
Systematic drug design: Gertrude Elion (1918–1999),
George Hitchings (1905–1998),
James Black (1924–2010)
G-protein coupled receptors:
Robert Lefkowitz (1943–) and
Brian Kobilka (1955–)

# Drug discovery & development

When searching for a new malaria treatment, Tu Youyou gathered hundreds of old recipes from traditional Chinese herbal medicine until she eventually isolated the **compound** artemisinin from a plant mentioned in one of these old texts. Since this discovery in 1972, artemisinin has treated many malaria patients.

Many new medications are no longer searched for in nature, but designed and tested in labs based on knowledge of **cell** signalling pathways and disease mechanisms.

Regardless of how they're initially discovered, all new medicines are tested in clinical trials to make sure they're safe and effective.

**WHAT COMES NEXT**
Discovering new drugs is only one way to create new medical treatments. Other methods include, for example, stem cell therapy (see p.174) or gene therapy (see p.180).

**SEE ALSO**
Vaccines, opposite
Protein structure, p.40
Cell signalling and transport, p.50
Bioinformatics, p.158
Antibiotics, p.165

# 93 | Transplants and blood transfusion

Grafting skin from one part of the body to another is one of the oldest methods of transplanting **tissue**. However, transplanting entire organs from one person to another was not successfully done until the 20th century.

It's not easy to find organs for transplants. While living people can donate liver tissue or a kidney, other organs can only come from the recently deceased.

It's much easier to donate blood, which can be stored in blood banks. Even though blood transfusions have been known since the 16th century, modern testing makes this process much safer.

**In 100 words**

If an organ is damaged, sometimes the only option is to replace it with an organ from a living or recently deceased healthy donor. To make sure that the immune system does not reject the donated organ, the recipient's blood should have a similar subtype of the human leukocyte **antigen** (part of the immune system) as the donor. They will also receive immunosuppressive **drugs**. People can also receive donated blood, usually via blood banks. Blood donors and recipients are matched by blood type, where A, B and Rhesus antigens determine which blood won't be rejected by the recipient's immune system.

## In 100 words

DNA fingerprints are created by using **restriction enzymes** or PCR to isolate or amplify sections of non-coding DNA called variable number tandem repeats (VNTRs). Each person has a unique combination of VNTRs of different length. If several VNTRs are isolated and separated by size using gel electrophoresis, it creates a pattern of bands on the **gel** that is different for everyone. This is a DNA fingerprint. DNA fingerprinting is used in forensics, paternity testing and medical **diagnostics**. It also works for plants and animals, so it can detect inbreeding in populations, which is useful for conservation projects or in agriculture.

# DNA fingerprinting

Unless you have an identical twin, your **DNA** does not match anyone else's in the world. The small differences between people are unique enough that they can be used to match a parent in a paternity test, find a criminal behind a crime scene, spot a genetic condition in an embryo or discover the level of genetic variability in a population.

Genetic fingerprinting uses gel electrophoresis to detect short segments of non-coding DNA which are repeated many times in the **genome**, but not the same number of times for everyone.

You inherit these repeated fragments from both parents, so DNA fingerprinting can show that a child shares some DNA with one parent and some with another.

**WHY IT MATTERS**
DNA fingerprinting is a quick method to test whether two DNA samples are from the same individual or a closely related family member.

**KEY THINKERS**
Alec Jeffreys (1950–)
Lalji Singh (1947–2017)

**WHAT COMES NEXT**
DNA fingerprinting only studies the length of some non-coding DNA fragments; it gives no information about the detailed DNA code. For that, we need Genetic sequencing (see p.156).

**SEE ALSO**

# IVF

When Louise Brown was born in 1978, she instantly made history as the world's first baby born via **in vitro fertilisation** (IVF) – sometimes referred to as a "test tube baby".

IVF has now been used millions of times to help people become parents. It is often used as infertility treatment, but the same technology can also help same-sex couples have children via donors or surrogates. It's also possible to screen the **zygote** for certain hereditary diseases such as cystic fibrosis or Huntington's disease before deciding to transfer it to the womb.

The process of IVF starts with hormone treatment before eggs are collected. Eggs are then fertilised in the lab and closely monitored before one egg is transferred to the uterus.

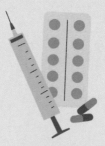

Hormone therapy for ovarian stimulation of egg production

Egg and sperm collection

In 100 words

IVF starts by collecting egg cells from ovaries. A hormone treatment increases the number of eggs that can be collected so that there is more than one to work with. These eggs are combined with sperm in a dish in a laboratory. This way, the first stages after fertilisation can be closely monitored and a zygote can be transferred into the womb after a few days. IVF is used in infertility treatment, surrogacy pregnancies (where another woman carries the embryo on behalf of the parents), pregnancies with donor sperm, or as a way to avoid passing on a genetic condition.

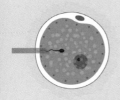

Fertilisation

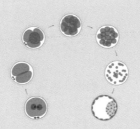

Embryos are cultured
for 5 days

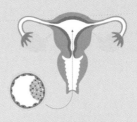

Embryo transfer
into uterus

# 96

# Biomarkers and genetic testing

**WHY IT MATTERS**
Measuring the right
biomarkers can show
whether someone is
likely to respond well
to a certain disease
treatment.

**KEY THINKERS**
Cholesterol as
biomarker:
John Gofman
(1918–2007)
BRCA1/2 in
personalised medicine:
Alan Ashworth (1960–)
Mary-Claire King
(1946–)
Olufunmilayo Olopade
(1957–)
ctDNA as biomarker:
Maurice Stroun
(1926–)

**WHAT COMES NEXT**
If a genetic test shows
that parents are
likely to pass on an
inheritable disease,
they can decide to use
IVF and select a zygote
that doesn't have the
genetic mutation
(see IVF, p.170).

**SEE ALSO**
Inheritance, p.86
Cancer, p.98
Mutation and
variation, p.42
DNA fingerprinting,
p.169

In 1955, John Gofman discovered that people who had higher levels of a particular form of cholesterol in their blood were more likely to get a heart attack. Since then, this low-density lipoprotein (LDL) cholesterol has been one of the most well-known biomarkers to predict heart disease.

A biomarker is something that can be measured in the body, such as hormone or **protein** levels, blood pressure, or a specific **genetic variant**. A recently discovered type of biomarker is circulating tumour **DNA** (ctDNA). These small DNA segments can be detected in blood to monitor how someone's cancer is progressing.

Some biomarkers help physicians to make decisions about **drug** treatments. This is called personalised medicine or precision medicine. For example, BRCA1 and BRCA2 gene variants are biomarkers that show which breast cancer patients are more likely to respond to a certain drug.

Genetic testing can also find out whether healthy people carry genetic variants that might put them at a higher risk of disease. This can be part of family planning, sometimes in combination with IVF. Some genetic tests are sold directly to anyone who is curious about their own DNA.

Biomarkers are measurements of the body, such as blood pressure or the level of a certain hormone, that give information about someone's current state of health or disease. Biomarkers can predict how a condition will progress or suggest the most suitable treatment. Personalised medicine provides medical care based on biomarkers. For example, people with a mutation in the BRCA genes are more likely to respond to certain breast cancer drugs. Genes can be biomarkers for disease, but genetic testing can also explore whether a healthy person inherited genetic variants that put them at a higher risk of developing a disease.

"We've found that patterns of gene activation vary in different cancers from different patients. The outcome ... depends on what's propelling the cancer."

**Olufunmilayo Olopade,** Oncologist

# 97

# Stem cell therapy

People with blood cancers such as leukaemia might be offered a bone marrow transplant. This is a form of stem cell therapy, because the donated bone marrow has haematopoietic stem **cells** that produce different types of healthy blood cells.

Adult stem cells, such as those in bone marrow, can only form a few types of cells. Embryonic stem cells can still become any cell of the body, so they would be even more useful for therapy, for example by making new tissues on demand. However, embryonic stem cells can only be collected from fertilised eggs that weren't implanted after an IVF procedure, so using them in therapy is controversial and, in some places, illegal.

Induced **pluripotent** stem cells (iPSCs) are an alternative to embryonic stem cells. IPSCs are created in a lab from specialised cells, such as skin or blood cells. These cells are reprogrammed to become stem cells again after adding certain transcription factors, called reprogramming factors. Researchers and medical doctors are currently studying how to make iPSCs safe to use in regenerative medicine.

Stem cells form specific cell types on demand, which can be used to repair tissues or replace cells that don't function well. This is regenerative medicine, which is used, for example, to treat cancers or neurodegenerative diseases such as Parkinson's disease. Bone marrow transfer is currently used to transplant haematopoietic stem cells in blood cancer treatment. Embryonic stem cells are very versatile because they can become many types of cell, but controversial to use in therapy. Instead, researchers are studying how to safely use induced pluripotent stem cells (iPSCs), which are made by reprogramming differentiated cells back into stem cells.

Induced pluripotent stem cells (iPSCs) are made by treating specialised cells with reprogramming factors. They can then differentiate into different cell types.

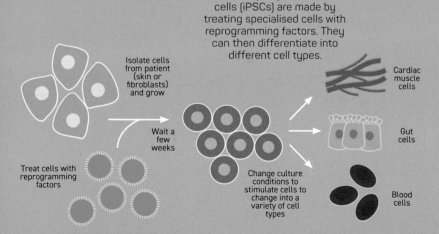

Isolate cells from patient (skin or fibroblasts) and grow

Wait a few weeks

Treat cells with reprogramming factors

Change culture conditions to stimulate cells to change into a variety of cell types

Cardiac muscle cells

Gut cells

Blood cells

# 98

# Cloning

**WHY IT MATTERS**
Molecular cloning is widely used to study the function of genes and proteins. Therapeutic cloning has the potential to be used in stem cell therapy.

**KEY THINKERS**
Molecular cloning:
Paul Berg
(1926–2023),
Herbert Boyer (1936–),
Stanley Cohen (1935–)
SCNT: John Gurdon
(1933–), Keith
Campbell (1954–2012),
Ian Wilmut
(1944–2023)

**WHAT COMES NEXT**
Reproductive cloning of animals could be a way to bring recently extinct animals back to life as a conservation effort (see Conservation, p.163).

**SEE ALSO**
Manipulating gene expression, p.152
Stem cell therapy, p.174

A sheep called Dolly was the first mammal to be successfully cloned in 1996 as a genetically identical copy of another sheep. This type of cloning, which produces an entire animal, is called reproductive cloning. It would be technically possible to clone humans in the same way, but there are currently many ethical, social and safety concerns. Reproductive cloning could also be used to bring back near-extinct or recently extinct animals.

Reproductive cloning is done with a method called **somatic** cell nuclear transfer (SCNT) which transfers the **nucleus** from a somatic **cell** (any cell besides an egg or sperm) into an egg cell that then develops into a blastocyst and is implanted in a womb.

The same method can also be used for therapeutic cloning by harvesting the blastocyst for embryonic stem cells instead of implanting it. However, the use of embryonic stem cells is itself also controversial.

A much more common and widely accepted use of cloning is gene cloning or molecular cloning. This involves making many genetically identical copies of a **gene** — for example by creating gene expression **vectors** in molecular biology research.

"Dolly the sheep told me that nuclear reprogramming is possible even in mammalian cells and encouraged me to start my own project."

**Shinya Yamanaka,** stem cell researcher

## In 100 words

Molecular cloning is the process of making many genetic copies of a gene — for example, by expressing it from a vector in bacteria.
It's also possible to clone entire **organisms**. For plants, this can be as simple as taking a cutting. Cloning animals is more difficult.
Mammals can be cloned using somatic cell nuclear transfer (SCNT), in which the nucleus of an egg cell is replaced with the nucleus of another cell. The egg develops into a blastocyst that can either be implanted into a womb to further develop (reproductive cloning) or be used to collect stem cells (therapeutic cloning).

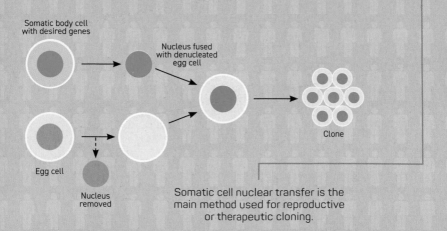

Somatic body cell with desired genes

Nucleus fused with denucleated egg cell

Clone

Egg cell

Nucleus removed

Somatic cell nuclear transfer is the main method used for reproductive or therapeutic cloning.

# 99

# Immunotherapy

The immune system is supposed to remove abnormal cells such as cancer cells, but cancers often manage to avoid detection. However, immunotherapy can make the body's immune system recognise cancer as a threat again. There are a number of different types of immunotherapy. Two of the most popular methods are checkpoint inhibitors and T **cell** transfer, but there are other immunotherapies that involve giving people **antibodies** or vaccines to attack the cancer cells.

T cell transfer therapy is also called adoptive cell therapy. It involves collecting T cell **lymphocytes** from a patient and returning them after they've been multiplied or improved in a lab. One popular method is to genetically edit the T cells so that they express a new **protein** called CAR, which is able to recognise cancer cells.

Immunotherapy is used to treat cancers and some autoimmune disorders, but could be used for other diseases in the future.

Immunotherapy is a medical treatment that helps the immune system fight off diseases such as cancer. Checkpoint inhibitors and CAR-T cells are some of the most-researched immunotherapies at the moment. Some cancer cells can switch off the immune response by interacting with checkpoint proteins on T cells. Checkpoint inhibitors are **drugs** that block this interaction, so that the immune system can fight off the cancer cells. CAR-T cells are made from regular T cells taken from a patient's or donor's blood. The cells are genetically edited to express a **receptor** that recognises cancer cells and then reinserted into the blood.

CAR-T therapy is a form of cancer immunotherapy where a patient's own T-cells are genetically modified so that they can recognise and destroy cancer cells.

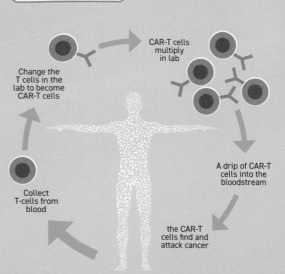

Change the T cells in the lab to become CAR-T cells

CAR-T cells multiply in lab

A drip of CAR-T cells into the bloodstream

the CAR-T cells find and attack cancer

Collect T-cells from blood

# 100

# Gene therapy

**WHY IT MATTERS**
Gene therapy has the
potential to treat many
diseases. A form of ex
vivo somatic cell gene
therapy is already in use
as CAR-T therapy, which
edits the body's T cells.

**KEY THINKERS**
French Anderson
(1936–)
Jean Bennett (c. 1955–)
Theodore Friedmann
(1935–)
Amit Nathwani (1960–)
And many others

**WHAT COMES NEXT**
Scientifically, it's
possible to edit human
DNA. But when biology
becomes medicine,
ethics and safety
are also important
factors (see Cloning,
p.176 and Stem cell
therapy, p.174).

**SEE ALSO**
Manipulating gene
expression, p.152
Genetic
sequencing, p.156
Immunotherapy, p.178

When the human **genome** was sequenced, it opened the
possibility of editing or adding human genes. This started
discussions and debates about the ethics of **gene** therapy.
Would people use **gene editing** in combination with IVF to
make designer babies? Would athletes change their **DNA**
to gain an advantage in competition?

Because of such concerns, gene therapy is tightly
regulated, and currently only a few types are available in
clinics. However, researchers continue to discover new
ways of gene editing, for example with CRISPR-Cas9, so
there will be other applications in the future.

"The impression sometimes created
among the public is that scientists are
working away in their labs, and
maybe they're not always thinking
about the implications of their work.
But we are."

**Jennifer Doudna,** biochemist

Researchers are studying whether
editing or adding human genes can be used
to treat disease. Gene therapy can be done in
the body (**in vivo**) or in cells that were removed
from the body (**ex vivo**) and then reinserted. If **germ cell** DNA
is edited, the changes are passed on to future generations.
This is illegal in many countries. Instead, most gene
therapy clinical trials focus on **somatic** cell gene
therapy, in which only non-germ cells are
edited. A lot of research is needed to
make gene therapies safe,
so it takes years for them
to become available
in clinics.

# Glossary

**Aerobic:** Using oxygen.

**Allele:** A version of a gene. The DNA sequence found in a specific location on a chromosome.

**Amino acid:** Smallest subunit of a protein. Twenty types of amino acids are common across all life forms.

**Anaerobic:** Not using oxygen.

**Anatomy:** Describing the layout of all organs and body parts.

**Antibody:** A protein produced by the immune system which recognises a very specific target antigen.

**Antigen:** A structural feature or molecule that is uniquely recognised by an antibody. For example, part of a protein on the surface of a pathogen.

**Apoptosis:** A natural process by which cells die when they are damaged or no longer needed.

**Atom:** The smallest unit of a molecule. Biologically important molecules are mostly made of the atoms carbon, hydrogen, oxygen, nitrogen, sulphur and phosphorus.

**ATP:** A molecule that stores energy. It's transported through the cell and releases energy where needed.

**Base:** The part of a nucleotide that is variable. Adenine (A), guanine (G), cytosine (C), thymine (T) or uracil (U).

**Biomass:** In ecology, biomass is the total mass (weight) of living organisms in an area.

**Cell division:** The process by which a cell splits into two new cells.

**Cell:** Smallest functional unit of all forms of life. All cells have a cell membrane, cytoplasm and organelles.

**Cellular transport:** Moving useful molecules or waste products within or between cells.

**Code:** A sequence of DNA or RNA defined by the order of the bases A, C, G, and T (or U).

**Codon:** Three nucleotides in mRNA forming a group that instructs which amino acid to add to a growing protein chain.

**Compound:** A molecule made of different types of atoms. Or used as adjective to describe insect eyes with many parts, each with their own lenses.

**Crystal:** A solid structure in which molecules form an organised and repeating pattern.

**Cytoplasm:** A gel-like substance that fills a cell.

**Decomposing:** Breaking down dead plants or animals into individual molecules.

**Development:** The process of forming and growing a new organism after reproduction takes place. Or, when discussing drug discovery and development, it's the process of testing potential new pharmaceutical drugs to find out if they can be used safely and effectively.

**Diagnostics:** Methods used to diagnose characteristics, symptoms or medical conditions.

**DNA:** Large molecule made of (usually) two strands of nucleotides with bases A, C, G and T.

**Drug:** A compound that has an effect on cells of the body, for example as medical treatment.

**Ecosystem:** A type of environment defined by the organisms that live there and their non-living surroundings.

**Enzyme:** Proteins that make chemical reactions in the cell go faster. Without enzymes, many reactions would be too slow for the cell to survive.

**Epigenetic:** Describing how gene expression is regulated by changes that are not in the genetic code itself.

**Epithelial:** Related to epithelium, a type of tissue that covers all inside and outside surfaces of the body — in, for example the skin or intestines.

**Ex vivo:** Using cells or tissues that were removed from the body.

**Exoskeleton:** A hard layer on the outside of certain animals, such as insects.

**Fertilisation:** Combining male and female gametes to form a zygote

**Fluorescent:** Producing light of a specific wavelength (colour) after absorbing light of another wavelength.

**Gamete:** Egg cells and sperm. Cells with only half of the typical number of chromosomes.

**Gel:** A material used in a lab to separate DNA, RNA or protein by size using an electric current.

**Gene:** A section of DNA that codes for a specific trait.

**Gene editing:** Purposely making changes to an organism's DNA.

**Genetic:** Related to genes or inherited via genes.

**Genetic diversity:** The overall genetic variation between all members of the same species.

**Genetic variant:** A part of a gene which is different between individuals.

**Genome:** All of the genes and non-coding DNA in an organism.

**Genotype:** The genetic variants or alleles in an individual.

**Germ:** A pathogen or microbe.

**Germ cell:** A reproductive cell that form gametes.

**Habitat:** The place where an organism lives, or its home within an ecosystem.

**Haemoglobin:** A protein in red blood cells which binds oxygen and transports it through the body.

**Hormones:** Molecules that travel through the body and regulate different processes when they interact with receptors in or on cells.

**Hydrophilic:** A description of molecules that are attracted to water.

**Hydrophobic:** A description of molecules that avoid water.

**In vitro:** Taking place in a dish in a lab.

**In vivo:** Taking place inside the body.

**Infectious disease:** A disease that is passed between individuals via a pathogen.

**Lipid bilayer:** Two sheets of phospholipids, layered with the hydrophilic phosphate groups on the outside and hydrophobic fatty acid chains on the inside.

**Lymphocytes:** Types of white blood cells with roles in the immune system, such as killing pathogens or producing antibodies.

**Membrane:** A lipid bilayer that separates the insides of cells or cell compartments from the outside.

**Microbe:** A micro-organism, such as bacteria or fungi.

**Molecule:** A chemical structure of two or more atoms bound together.

**mRNA:** Messenger RNA that is created as a copy (transcript) of a gene from DNA and can be translated into a protein.

**Myosin:** A type of protein that uses ATP to move, like a "molecular motor". It's involved in muscle contraction and in cellular transport.

**Neuron:** A cell of the nervous system, or nerve cell, that has long extensions to connect to other cells.

**Nucleotide:** The smallest subunit of DNA or RNA, made of a phosphate, sugar and a base.

**Nucleus:** An enclosed area within a cell where DNA is stored and transcribed to mRNA.

**Nutrients:** Molecules that provide organisms with energy or building blocks needed to survive and grow.

**Oncogene:** A gene that causes cancer when it has a genetic mutation. Before the mutation happens it is a proto-oncogene.

**Organelle:** A specialised compartment within a cell.

**Organism:** An individual life form made of one or more cells.

**Pathogen:** An organism that causes disease, such as a virus or bacteria.

**Phenotype:** An observed trait. The visual or measurable output of a genotype.

**Photosynthesis:** A process used by plants and algae to convert sunlight and carbon dioxide into sugars and oxygen.

**Physiology:** Describing how an organism's body functions.

**Pluripotent:** For a stem cell to have the ability to become any type of cells in the body.

**Pollination:** Transferring pollen from one flower to another to fertilise plants. Often carried out by insects such as bees.

**Polymerase:** An enzyme that grows a DNA or RNA strand by adding nucleotides.

**Population:** All members of a species within a certain area.

**Protein synthesis:** The process of forming a new protein by adding amino acids to a growing chain.

**Protein:** Large molecule formed by a chain of different amino acids folded into a 3-D structure. End product of a gene.

**Receptor:** A protein that passes on a signal in a cell when it's bound to a molecule, such as a hormone. Transmembrane receptors communicate information from outside of the cell to the inside.

**Recombinant:** Having DNA from different sources — for example, by adding a vector with a new piece of DNA to a cell.

**Reduced:** Having an additional hydrogen atom.

**Reproduction:** The process of producing offspring. Sexual reproduction involves fertilisation, by which two gametes form a zygote. Asexual reproduction only needs cell division.

**Respiration:** Cellular respiration is a series of chemical processes that produce energy for the cell by breaking down nutrients.

**Restriction enzyme:** An enzyme that cuts a piece of DNA at a specific short sequence of bases.

**RNA:** Large molecule made of either one or two strands of nucleotides with bases A, C, G and U.

**Sample:** A small piece or section of a larger unit that represents all of it. Used for testing or measuring.

**Scan:** An image created by gradually measuring one section at a time.

**Sequence:** The order of nucleotides in DNA and RNA, or the order of amino acids in a protein.

**Sequencing:** Finding the sequence of a piece of DNA or RNA.

**Single-celled:** Consisting of just one cell.

**Somatic:** Related to cells of the body other than the gametes or germ cells.

**Species:** A group of similar organisms

that can breed with each other. It's the most specific taxonomic rank.

**Strand:** One chain of a DNA or RNA molecule. Double strands are connected by matching base pairs A-T (or A-U) and C-G

**Taxonomy:** A method of grouping all life forms into categories or ranks.

**Telomere:** The tips of chromosomes

**Tissue:** A structure in the body made of cells with a similar function.

**Trait:** A characteristic of an organism.

**Vector:** A circular piece of DNA or RNA that is used in a lab to bring new genetic material into a cell — for example, to add a gene or inhibit expression of an existing gene.

**Zygote:** A fertilised egg cell. The first stage of embryonic development.

# Index

# About the author

Eva Amsen is an award-winning writer and science communicator, focused on biology, chemistry, the culture of science and the common ground between science and the arts. Her writing has been published widely, including in *Nature*, *Nautilus*, *The Scientist*, *Undark Magazine*, *PNAS*, *Spacing Magazine*, *Hakai Magazine*, Forbes.com, BOLD blog and in the books *The Open Laboratory 2009* and *The Best Science Writing Online 2012*. Eva has a PhD in biochemistry from the University of Toronto.

Eva would like to thank Shelley Edmunds, Åsa Karlström, Kelly Reimer and Maartje Winkler for feedback on some of the topics and definitions in this book.

## In association with the Science Museum

The Science Museum is part of the Science Museum Group, the world's leading group of science museums that share a world-class collection providing an enduring record of scientific, technological and medical achievements from across the globe. Over the last century the Science Museum, the home of human ingenuity, has grown in scale and scope, inspiring visitors with exhibitions covering topics as diverse as robots, code-breaking, cosmonauts and superbugs. www.sciencemuseum.org.uk.

## Picture Credits

The publisher would like to thank the following for their kind permission to reproduce their photographs:

(Key: a-above; b-below/bottom; c-centre; f-far; l-left; r-right; t-top)

**1 Shutterstock.com:** Olga Bolbot. **2–3 Dreamstime.com:** Metelsky25. **4–5 Shutterstock.com:** Vectores de Lia. **8–9 Shutterstock.com:** Noval A Kahfi. **15 Shutterstock.com:** Natasha Barsova. **19 Science Photo Library:** Paul D Stewart (br). **20–21 Dreamstime.com:** Patrick Guenette. **22–23 Shutterstock.com:** Svetla. **23 Shutterstock.com:** Onica Alexandru Sergiu (b). **24–25 Shutterstock.com:** Maxim Gaigul. **27 Dreamstime.com:** Ldarin (br). **31 Shutterstock.com:** ShadeDesign (br). **33 Dreamstime.com:** Sergei Sizkov (cb). **Shutterstock.com:** In-Finity (cl); ShadeDesign (br). **41 Dreamstime.com:** VectorMine (cb). **58–59 Shutterstock.com:** Curly Pat. **59 Dreamstime.com:** VectorMine (b). **60–61 Shutterstock.com:** Olgastocker. **65 Shutterstock.com:** Mallinka1 (b). **68 Shutterstock.com:** AngryBrush. **71 Dreamstime.com:** Metelsky25. **73 Dreamstime.com:** Designua (cb). **75 123RF.com:** pro100vector (bc). **76–77 Shutterstock.com:** Vectores de Lia. **78–79 Dreamstime.com:** Natallia Yatskova. **81 Shutterstock.com:** In-Finity (crb). **82–83 Dreamstime.com:** Luayana (b). **90–91 Dreamstime.com:** Sergei Sizkov. **93 Shutterstock.com:** In-Finity (cb). **100–101 Dreamstime.com:** PHHY (b). **102–103 Shutterstock.com:** Shafran. **104–105 Dreamstime.com:** Eveleen007. **107 Shutterstock.com:** SilverCircle. **108 Shutterstock.com:** Saylee Rampurikar (cb). **110–111 Shutterstock.com:** Moh. Idrus. **112 Shutterstock.com:** Olga Bolbot (cb). **113 Shutterstock.com:** okili77 (cb). **117 Shutterstock.com:** Saylee Rampurikar (br). **118–119 Shutterstock.com:** Krzepax. **123 Dreamstime.com:** Alexander Pokusay (b). **124 Dreamstime.com:** Alexander Pokusay (crb). **125 Dreamstime.com:** VectorMine (cb). **126–127 123RF.com:** Watchara Khamphonsaeng. **127 Shutterstock.com:** AtlasbyAtlas Studio (cb). **129 Shutterstock.com:** Olga Bolbot (clb). **130–131 Shutterstock.com:** afry_harvy. **132–133 Shutterstock.com:** Tiberiu Stan. **135 Shutterstock.com:** Brian Goff. **140–141 Shutterstock.com:** Olga Bolbot. **144–145 Shutterstock.com:** Protasov AN. **145 Shutterstock.com:** Protasov AN (b). **146 Alamy Stock Photo: Photo Researchers / Science History Images (crb). 151 Shutterstock.com**: Your MUSE (br). **156–157 Science Photo Library:** Tek Image. **158–159 Shutterstock.com:** Foryoui3. **160–161 Shutterstock.com:** Granit_Fan. **165 Dreamstime.com:** Ikorch007. **170–171 Shutterstock.com:** Pikovit (cb). **172–173 Shutterstock.com:** natrot. **176–177 Shutterstock.com:** KostiantynL. **179 123RF.com:** pro100vector (b)

Cover images: Front: **Alamy Stock Photo:** funkyfood London - Paul Williams cr; **Dreamstime.com:** Katerynakon br; **Getty Images:** Kateryna Kon / **Science Photo Library tr; Spine:** Dreamstime.com

All other images © Dorling Kindersley

**DK LONDON**
**Editors** Florence Ward & Millie Acers
**Senior Designer** Anna Formanek
**Senior Acquisitions Editor** Pete Jorgensen
**Managing Art Editor** Jo Connor
**Production Editor** Siu Yin Chan
**Production Controller** Louise Minihane
**Managing Director** Mark Searle

**Written by** Eva Amsen
**Designer** Neal Cobourne
**Jacket Designer** Steven Marsden

DK would like to thank Caroline Orr for copyediting, Caroline Curtis for proofreading, Emma Caddy for indexing and Izzy Merry for design assistance.

First American Edition, 2024
Published in the United States by DK Publishing
1745 Broadway, 20th Floor, New York, NY 10019

A catalog record for this book is available from the Library of Congress.
ISBN 978-0-7440-9378-0

Printed and bound in China

In association with
Science Museum
Exhibition Road, London SW7 2DD
www.sciencemuseum.org.uk

Every purchase supports the museum.

**www.dk.com**

DK books are available at special discounts when purchased in bulk for sales promotions, premiums, fund-raising, or educational use.
For details, contact: DK Publishing Special Markets,
1745 Broadway, 20th Floor, New York, NY 10019
SpecialSales@dk.com